Scala Programming for Big Data Analytics

Get Started With Big Data Analytics Using Apache Spark

Irfan Elahi

Apress®

Scala Programming for Big Data Analytics

Irfan Elahi
Notting Hill, VIC, Australia

ISBN-13 (pbk): 978-1-4842-4809-6   ISBN-13 (electronic): 978-1-4842-4810-2
https://doi.org/10.1007/978-1-4842-4810-2

Managing Director, Apress Media LLC: Welmoed Spahr
Acquisitions Editor: Celestin Suresh John
Development Editor: Matthew Moodie
Coordinating Editor: Aditee Mirashi

Cover designed by eStudioCalamar

Cover image designed by Freepik (www.freepik.com)

Distributed to the book trade worldwide by Springer Science+Business Media New York, 233 Spring Street, 6th Floor, New York, NY 10013. Phone 1-800-SPRINGER, fax (201) 348-4505, e-mail orders-ny@springer-sbm.com, or visit www.springeronline.com. Apress Media, LLC is a California LLC and the sole member (owner) is Springer Science + Business Media Finance Inc (SSBM Finance Inc). SSBM Finance Inc is a **Delaware** corporation.

For information on translations, please e-mail rights@apress.com, or visit http://www.apress.com/rights-permissions.

Apress titles may be purchased in bulk for academic, corporate, or promotional use. eBook versions and licenses are also available for most titles. For more information, reference our Print and eBook Bulk Sales web page at http://www.apress.com/bulk-sales.

Any source code or other supplementary material referenced by the author in this book is available to readers on GitHub via the book's product page, located at www.apress.com/978-1-4842-4809-6. For more detailed information, please visit http://www.apress.com/source-code.

Printed on acid-free paper

Dedicated to my parents and wife who trusted and supported me all the way to becoming who I am today

Table of Contents

About the Author

Irfan Elahi is currently working in Deloitte, Australia and specializes in Big Data and machine learning. He possesses years of diverse experience in end-to-end lifecycle and in designing, developing, and deploying production-grade Big Data and analytics solution architectures in leading cloud environments (Azure, AWS, and GCP) that support a wide array of business use cases (including data lake, scalable predictive and graph analytics, and stream processing, to name a few). His experience extends to DevOps, platform governance, and administration aspects of Big Data analytics. In addition to his professional achievements, Irfan presented at the *DataWorks Summit* in Sydney in 2017 about in-memory Big Data technologies and at a number of universities and meetups all around the world. He also launched Udemy courses on Apache Spark for Big Data analytics and R programming for Data Science, teaching thousands of students from over 150 countries.

About the Technical Reviewer

Manoj has served the software industry for 19 years. He holds an engineering degree from COEP, Pune (India) and has enjoyed his exciting IT journey.

Being a Principal Architect at TatvaSoft, Manoj has taken many initiatives in the organization, ranging from training and menztoring the teams, leading the Data Science and ML practice, to successfully designing client solutions from different functional domains.

Starting as a Java programmer, he was fortunate to have worked on multiple frameworks with multiple languages and can claim to be a full stack developer. In the last five years, he has extensively worked in the field of BI, Big Data, and machine learning with technologies like Hitachi Vantara (Pentaho), Hadoop ecosystem, TensorFlow, Python-based libraries, and so on.

He is passionate about learning new technologies and trends and reviewing books. When he's not working, he's either working out or reading/listening to infinitheism literature. Manoj would like to thank Apress for giving him the opportunity to review this title and his two daughters, Ayushee and Ananyaa, for their understanding during the process.

Acknowledgments

The fact that there are generally many people in your life that have direct or indirect influence on your success cements the notion that one can't do complete justice in the acknowledgements section found in books. Still, I'll endeavour in the best of my capacity and I apologize in advance to those I didn't explicitly happen to mention; rest assured your efforts/ contributions are duly acknowledged.

Moral support in each step of the book writing process fuels one's motivation and dedication to achieve the elephantine milestone of writing and completing a book. In this context, I can't emphasize enough the moral support from my parents and wife who were always there for me. Their encouragement and trust were a true driving force. My gratefulness additionally extends to my workplace friends, Shahban Riaz and Fahad Sohail, who shared their feedback from their broader perspectives about how to align this book to resonate better with the target audience.

Transforming your ideas from a nascent state into a book is an involved process but it was streamlined significantly courtesy of the amazing support from Apress. I am grateful to Celestin Suresh John whose confidence in me and my book's proposition led to its tangible realization. Aditee Mirashi has been a wonderful editor to work with and her cooperation in this process can't be appreciated enough. Thanks to Matthew Moodie for the investment of his time and effort to thoroughly scrutinize my book and identify typos and opportunities for improvement. A huge credit goes to Manoj Patil for his dedication to validating technical aspects of the book. The collective experience of working with Apress has been stupendous indeed.

ACKNOWLEDGMENTS

This section would be incomplete if I didn't acknowledge the people instrumental in my journey to excellence (still a long way to go indeed) in big data and analytics. My deepest gratitude to my elder brother, whose support in the initial days of my career paved my path to this domain. Lastly, a huge shout-out to my current employer, Deloitte, which enabled me to apply (and enhance) my skill-set by working on challenging, real-world projects. Huge thanks goes to Murad Khan (and other Deloitte Partners) for trusting my abilities to work in several strategic engagements in big data and analytics.

Foreword

As a leader in one of Australia's largest professional services firm, I often get credit for introducing talented individuals to the firm. While that may be true in a lot of cases, I do not correct people when they assume I am responsible for Irfan as well. That is simply because not only is Irfan one of the most dedicated professionals I've come across in my career, but also that he has an uncanny ability to resolve every challenge thrown his way. This book, in a similar fashion, tries to address all dimensions of Scala programming for Big Data analytics. It is worth every page because it is based on real-life problems that Irfan has solved over the years while working in a series of engagements in the firm. I am certain that this book is imbibed with Irfan's unique pedagogy, coupled with his emphasis on best practices and attention to detail, and thus will be a significant value for the readers.

Murad Khan

Partner | Consulting | Analytics & Cognitive

Deloitte Touche Tohmatsu, Australia

Introduction

Let me start by congratulating you on your decision to read this book.

I understand that with this decision, there will be significant magnitude of hopes and expectations that you will have from this book. This introduction sets the context and expectations so that we are on the same page, which will lead to the best learning process.

First, as evident from the book's title and description, this book is about Scala. You will learn the foundations of Scala, which is one of the hottest and in-demand programming languages out there. With the rise of Big Data platforms like Apache Spark and Kafka, its demand has skyrocketed further. So your decision to learn Scala will definitely be fruitful, provided you put the effort needed to learn it. With that being said, the book's title outlines the scope of this book quite clearly: *Scala Programming for Big Data Analytics*.

Scala is a general-purpose language and can be used for a number of use cases like Big Data development, web applications development, and numerical computation, to name a few. This book was developed with a focus on Big Data development. With that being said, the book doesn't teach you about Big Data Development in detail. Rather, it covers and teaches the concepts of the Scala language that are relevant to getting started with Big Data development.

Now you may naturally ask, is it really important? And why this book? Let me introduce myself quickly to clarify the point. I have been working in Big Data and machine learning for years and I am proud to confess that I am a self-taught engineer/data scientist. I learned all of the technologies, frameworks, literature, tools, etc. that are used in Big Data on my own. So I have strong empathy for those who want to start in this domain and understand the challenges they may face, as I have faced similar challenges in my career. Big Data and machine learning (or Data Science in general) are huge domains and one can easily get overwhelmed with so many things that need to be learned to develop excellence. Specifically for Big Data development, I've seen people who either don't have a computer science background or, even when they do, they find one a critical block: They aren't skilled in the languages that are considered standard in the Big Data landscape. Hadoop, one of the de facto platform/technologies that powers Big Data technology, is primarily developed in Java. Many of the recent and heavily used Big Data technologies, like Apache Spark and Kafka, have been developed in Scala.

Apache Spark, specifically, is one of the most widely used Big Data processing frameworks used for a number of use cases (you'll find some details about this in subsequent chapters), and it is developed in Scala. If you want to use it for your projects/use cases, you are expected to learn Scala. That's where the challenge happens. Big Data engineers struggle a lot in developing the skillset in Scala, which is a prerequisite to use Apache Spark. They follow books that are too detailed and voluminous and cover Scala in too much depth; many of those concepts generally aren't critical for getting started in Apache Spark development. It's for this very reason that Cloudera, one of the leading Hadoop commercial vendors, offers "Just Enough Scala" training for all those who want to learn and enroll in their Apache Spark and Hadoop Development certification program. If you can afford to enroll in their expensive training, please do. Or if you want to learn just enough of Scala from a self-taught Big Data engineer who is currently working in the industry and has been using Scala for Big Data and general-purpose application development, this is the exact opportunity this book provides. You can leverage my learning and exposure that I have shared in this book and become skilled in Scala in a focused and no-nonsense way. If you ask me, nothing can rival this value.

With that being said, let me just quickly point out a couple of nuances in this book. This book covers Scala concepts related to getting started with Apache Spark development. Even those concepts have depth of their own. I will cover those concepts to the required level of depth along with a lot of examples that you can use for hands-on practice. Additionally, I will highlight what you need to study on your own as well. You may find this style to be different than books that are meant to cover everything, but if I start going into details of those auxiliary concepts, the book will digress from the main focus—which is to get you up to speed with Scala so that you can focus on learning Apache Spark.

So you will find me suggesting that you research specific topics and concepts. I have gained this inspiration from attending training programs of many leading companies, like Cloudera and Microsoft, which give you a problem and expect you to sort out on your own without relying on the instructor to teach you. Those training programs have proven to be highly effective. Similarly in this book, if you follow the pointers that I highlight, it will be helpful as it will broaden and deepen your level of skills. If you don't, you'll still get a lot from this book. It's all about how much you want to invest in developing skills. Thus, without further ado, let's learn Scala, shall we? I hope that you will find my style of teaching to be engaging and interesting, as I approach pedagogy in a unique way. That's the reason that I am Data Science trainer in Deloitte, have spoken at a number of forums like universities, global summits, and meetups, and have thousands of students enrolled in my Udemy courses.

Best of luck and let the journey to excellence begin!

CHAPTER 1

Scala Language

Programming languages have been around for a very long time and have evolved significantly. Starting from the very foundational binary language, which merely consisted of 0 and 1 bits, a huge array of languages has been developed over the years to address a number of growing challenges in different contexts. All of these languages have a core purpose—to enable developers to write instructions in a way that is understandable by computers with the goal of completing a specific task. Some languages specialize in a specific set of tasks, such as developing web applications, whereas others are preferred for Data Science and machine learning tasks. Yet others are recommended for developing applications on the Windows operating system, and the list goes on.

Languages are generally placed into different categories, which assists in developing some understanding of them:

- High level or low level—Relates to abstraction, i.e., how close your instructions correspond to what computers actually do.

- Object-oriented—Relates to the design of programs consisting of objects, their properties, and their functionality, along with their interactions with each other.

- Static or dynamic—Relates to the strength of the association of the type to the objects defined.

© Irfan Elahi 2019
I. Elahi, *Scala Programming for Big Data Analytics*,
https://doi.org/10.1007/978-1-4842-4810-2_1

Note   *Type* is a commonly used term in programming language literature. It refers to data types of information that you use and create in your program. For example, if you want to use numbers, there are types for that, for instance, `Integer`. If you want to write alphanumeric characters such as "`Samsung Galaxy S7`", you use the `String` type.

- Functional programming—Relates to the level of significance that functions hold in a language, i.e., are they treated the same way as other objects or not (e.g., assigning to variables, passing as arguments to functions, immutability of variables, etc.).

Another way to understand the notion of programming languages is that some languages are derivatives of other languages. For example, Scala, as you will find, is a superset/descendent of Java language. Scala code gets compiled to a Java-specific format called *Java bytecode* and runs in the Java Virtual Machine environment. Python's code, even though it has many implementations like CPython and Jython, is compiled in the C language.

In a nutshell and to keep matters simple, languages allow you to interact with computer systems. They allow you to specify instructions that computers can understand and process, which ultimately translates to addressing and solving many problems, ranging from simple mathematical calculations to complex tasks like processing data on different systems in a distributed and parallel manner.

You can easily find a lot of general information about programming languages, so I won't cover the general aspects here. Rather I'll keep this chapter short and will highlight some key interesting characteristics of the Scala programming language specifically in the context of Big Data analytics and Apache Spark development.

Getting to Know Scala

Scala, which is a short form of SCalable LAnguage, originated from 'École Polytechnique Fédérale de Lausanne' (EPFL), Switzerland, in 2003, with the goal of realizing a high-performance and highly concurrent language that combines the strength of the following two leading programming patterns on the Java Virtual Machine (JVM) platform:

- Object-oriented programming

- Functional programming

Both of these patterns allow you to express the problem at hand in an efficient and reusable way. Object-oriented principles focus on building objects and their interactions with other objects. Functional programming concentrates on functions being prime objects in programming, immutability of data (i.e., inability to change a variable's state), purity of functions (i.e., whether functions can change a value beyond their scope), and making iterations more implicit. Many of these concepts will become clear as you progress in this book.

Also, it is an established fact (Source: `https://www.scala-lang.org/old/node/3069`) that using Scala leads to increased developer productivity, as your code becomes better (because you use more immutable structures, thus reducing a lot of side-effects), simpler and more expressive, and statically typed (the data type of a variable is strongly adhered to). In other languages, such as Python, if you define a variable, you can store a number or a string value in it. However, if a language such as Scala is statically typed, you can't store a number value in a string variable because doing so results in an error. This feature helps a lot when you deploy such applications in production where the application processes data. You'll see more about static typing later in this chapter.

Why Learn Scala?

Apart from the fact that Scala is the *lingua franca* of Big Data development, particularly when you use the leading distributed computation framework Apache Spark, and that it leads to better programs and happy developers— there are a lot of other perks to learning Scala:

- Programmers skilled in Scala are considered highly valuable in the marketplace all around the world and earn highly competitive salaries (Source: `https://adtmag.com/articles/2017/08/18/go-scala-salaries.aspx`).

- It's heavily being used in the industry in companies of all scales, including Netflix, LinkedIn, and Twitter (Source: `https://techcrunch.com/2016/06/14/scala-is-the-new-golden-child/`).

- It has been featured prominently in developers' surveys conducted all around the world, indicating strong interest in this programming language by the global development community (Source: `https://jaxenter.com/survey-results-top-programming-languages-131820.html`).

Scala and Java

If you don't know already, Java is one of the most famous and widely used programming languages out there. More than a billion devices use Java. One of many reasons for Java's success is that it provides platform independence, i.e., you can write Java code once and run it on any platform, like Windows/Linux.

Additionally, if you are inclined to learn about Big Data, you may have heard about Hadoop, which is the de facto framework that powers Big Data platforms. Hadoop, which actually is a suite of services, is primarily written in Java.

I suggest you develop some understanding of Java, the Java Runtime Environment (JRE), the Java Development Kit (JDK), the Java Virtual Machine (JVM), and Java bytecode. These concepts will help you in the long run if you intend to develop an extended skillset in Scala.

How Does Scala Relate to Java?

Scala is referred to as a JVM (Java Virtual Machine) language, which means that when you compile Scala code, it gets compiled to Java bytecode. Java bytecode is an abstract machine language or instruction set that is executed by the JVM. Think of the JVM as a program that executes other programs (in the context of Java, it runs Java code). It's like a virtual environment that runs on top of any operating system (thus providing platform independence) and manages system resources. So, a Scala program gets compiled to Java bytecode, which then runs in the JVM.

Although there are similarities between Scala and Java, there are differences between the two as well. In a way, Scala tries to address many of the shortcomings of Java. One of those is Java's verbosity, which is elaborated further in the following sections in the chapter.

Interoperability with Java Libraries

As Scala is a JVM language, this is itself an interesting feature. You can use the Java language's libraries in Scala! This is a great perk, as Java has a huge ecosystem of libraries for a variety of purposes and you can use them in your Scala program. Scala's own libraries are great and are growing as well, but being able to use Java libraries adds a lot to the usability and diversity of Scala.

Scala and Java Verbosity

There is one interesting difference between the two languages related to verbosity. Verbosity refers to how much code you write to accomplish a task. Java is notorious for being too verbose, i.e., you have to write a lot of code even to do a simple task like print a message onscreen. On the other hand, Scala's verbosity is quite low. Additionally, as Scala is a functional programming language, one of the benefits is that it allows you to express many operations (e.g., loops) in a succinct way without explicitly writing looping statements. If this isn't making much sense right now, don't worry as you will be doing plenty of this stuff later in this book.

This example provides a nice comparison of the verbosity of Java and Scala. Here's the code in Java first:

```
public class Person {
    private final String name;
    private final double age;
    public Person(String name, double providedAge) {
        this.name = name;
        this.age = providedAge;
    }
    @Override
    public int hashCode() {
        int hash = 10;
        hash = 23 * hash + Objects.hashCode(this.name);
        return hash;
    }
    @Override
    public boolean equals(Object obj) {
        if (obj == null) {
            return false;
        }
```

```
    if (getClass() != obj.getClass()) {
        return false;
    }
    final Test other = (Test) obj;
    if (!Objects.equals(this.name, other.name)) {
        return false;
    }
    if (Double.doubleToLongBits(this.age) != Double.
    doubleToLongBits(other.age)) {
        return false;
    }
    return true;
}
@Override
public String toString() {
    return "Test{" + "name=" + name + ", age=" + age + '}';
}
}
```

The same code in Scala looks like this:

```
case class Person(name: String, age: double)
```

All of these benefits translate to increased development productivity
and lower code maintenance cost.

Scala: A Statically Typed Language

One of the dimensions in which languages are compared is whether a
language is statically typed or not. What this basically means is that if a
language is statically typed, it has a type system that is checked at compile
time. For example, if you write a module/function in your program that

expects input to be of a specific type, upon compilation, it will be checked for that condition. It also means that the type is associated/bound to a variable. When you define a variable, its type is defined and it can only store values of that type.

On the other hand, with languages that are dynamically or loosely typed, such checks are performed at runtime (when the program is actually run). This poses a challenge because it's quite risky to deal with errors at runtime. It's better to identify errors at compile time, i.e., during compilation prior to execution.

Scala falls under the category of a statically typed language and so does Java. Some examples of loosely typed language include Lua and Python.

Apache Spark and Scala

The rise of Scala has strong roots in the Big Data phenomena. Big Data, a term that has many interpretations in different contexts, primarily means that if you can't process data in a single machine, you resort to a cluster of machines that are interconnected. Those cluster of machines work in concert and perform processing and computation in a distributed manner. This is the new and de facto paradigm of scalable computing. Previously, the approach was to make an individual server more powerful by increasing its resources; however, one can only scale an individual machine to a finite extent. Whereas you can scale a cluster of machines as much as needed, per your processing requirements.

In the world of Big Data, you will often hear the term "Hadoop". Hadoop, as alluded to, is a suite/collection of services and software that work in conjunction to enable Big Data computation and storage. There is so much to talk about Hadoop, but to keep things focused, I'll just highlight one specific aspect of it. Services in Hadoop can be majorly placed into two categories: *storage* and *compute*. To give you a taste of Hadoop, but to keep it succinct, I'll just cover compute here.

In the compute category of Hadoop, there are a lot of services/tools/ frameworks that allow you to perform processing in parallel on a cluster of machines. Hadoop originally started with the MapReduce framework and then many other engines spawned, including Apache Spark, Apache Storm, Apache Flink, Apache Impala, and so on. The Apache Spark framework got the most traction and has become the go-to choice for performing a number of Big Data and analytics tasks, such as ETL (extraction, transformation, and loading), machine learning, and graph analytics, to name a few.

And guess what: Apache Spark is written in the Scala language! If you want to use Apache Spark, you use its APIs (application programming interface, which are the libraries that allow you to interact with the system) available in Scala. Apache Spark does have API support for other languages, including Python, Java, and R, but this support is not as robust as Scala's.

Let's spend some moments getting to know Apache Spark.

Apache Spark, among other capabilities, provides in-memory computation. Previously, compute engines like MapReduce required a lot of disk I/O to perform their tasks. Usually, the workflow of tasks in these Big Data frameworks consists of different stages. For instance, if you want to determine the frequency of words in a collection of documents, you generally perform these series of tasks:

1. Load the documents.

2. Tokenize the words, i.e. convert each line of the document into words.

3. Transform each word into a different form that may look something like (word,1) in one structure.

4. Combine the same words and add the number portion together.

By doing so, you can count the frequency of the words in a collection of documents. As you can see, we have mapped this into a series of phases/ stages. If you use MapReduce, each of these intermediate stage's result is written to disk and then read by subsequent stages.

That's where issues happen: disk I/O operations are always expensive. By *expensive*, it means that they take a significant amount of time to complete. If you compare them with RAM, the operations there are much less expensive. It's pretty fast to write/read from RAM. This is exactly the leverage that Spark maximizes and provides via its in-memory computation capability. The intermediate results aren't written to disk and are rather performed while staying in RAM (more specifically, the on-heap memory of the JVM). This results in much better performance; the literature states that it's 100 times faster than MapReduce (Source: `https://dzone.com/articles/apache-spark-introduction-and-its-comparison-to-ma`). Figure 1-1 highlights the difference between Hadoop MapReduce and Apache Spark.

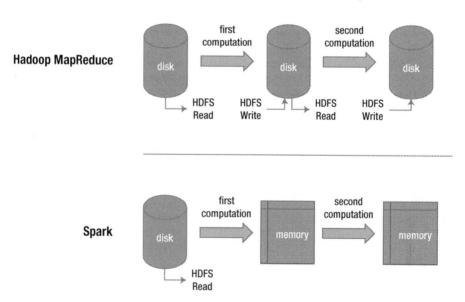

Figure 1-1. *Comparison of input/output (I/O) operations in Hadoop MapReduce and Apache Spark, which is one of the reasons that Apache Spark is significantly faster than MapReduce*

It's mainly because of this that Apache Spark is heavily used in businesses all around the world. It's the most active open source Apache project and the community is working extensively to strengthen it over time.

As Apache Spark is written in Scala, this places Scala in a special and favorable position. Scala has become the language of choice whenever you want to do anything with Apache Spark. The reason behind such preference is manifold. First, Apache Spark is evolving with new releases and versions coming on a periodic basis. All of those new language features that come with new releases are initially made available to its Scala language API. Although Spark APIs are available in other languages as well, for instance Python and R, new features are supported only in the Scala APIs.

Secondly, as Apache Spark is open source, developers and businesses access the source code and make modifications to it when the default implementation of Spark doesn't have what they need.

It always helps to understand what each function or class does. To develop such an understanding of the Apache Spark classes and functions, you need to have a strong skillset in Scala.

Apache Spark isn't the only well known project in the Big Data landscape built in Scala. Many other famous technologies, like Apache Kafka and Apache Samza, are also built in Scala.

Scala's Performance Benefits

A lot can be said and debated about this topic, but the fact remains—you get the best performance in Apache Spark when you use Scala. Databricks, a commercial Big Data vendor and the founder of Apache Spark, conducted an extensive comparison of performance when Spark is used in Python versus Scala. The results are available at the following link:

```
https://databricks.com/blog/2015/04/24/recent-performance-
improvements-in-apache-spark-sql-python-dataframes-and-
more.html
```

This research clearly indicates that when RDDs (the core abstraction of Apache Spark; for the time being, just think of them as collection-like arrays) are used in Apache Spark APIs that are available in both Scala and Python, the performance difference between the two languages becomes significant. Apache Spark APIs have graduated to dataframes (another data structure that Spark provides) and when dataframes are used, the performance difference becomes negligible. However, RDDs are still used extensively and thus can have a substantial impact on the performance of your Apache Spark applications.

There are a lot of technical reasons behind why JVM languages are better than interpreted languages in terms of speed, but let's not go down that path. For you, it's more than enough to understand that with Scala, you are assured to have the best level of performance (provided, obviously, you have written the code correctly).

Learning Apache Spark

Although this book is about learning Scala, the natural next step after learning Scala is to develop your skillset in Apache Spark. If you want to learn from my best-selling course on Udemy, which has been the highest-rated course multiple times, you can enroll in the course by visiting the following link: `https://www.udemy.com/apache-spark-hands-on-course-big-data-analytics/`.

EXERCISES

- Learn about the Java language and its associated terms, such as JDK, JRE, bytecode, and JVM.

- Learn why Scala was developed.

- Research more about Apache Spark.

- Understand the different use cases for Big Data.

- Understand the various applications of Scala.

- Research the other famous products developed in Scala, and if possible, why?

- Research concerns about the performance of certain languages, such as Python.

CHAPTER 2

Installing Scala

Before you can use Scala to write exciting programs and make your way to excellence in Apache Spark, you need to install Scala in your system. Even if Scala isn't installed on your system, you can open Notepad or any text editor of your choice and write Scala code/expressions and then save the contents of that file with a .scala extension. However, that won't help you in achieving your desired goal, which is to compile the program that you wrote, run your Scala program, or package it (mostly in the form of JAR files). Not to mention that you won't be using the facilities that come with the Scala shell (more on that later). These and many other characteristics are enabled if you have Scala installed on your system.

Just a note that there are a number of online tools available, including Scastie (`https://scastie.scala-lang.org`) and Databricks (which we will use in later chapters), but it's always productive to have Scala accessible even when you are offline. Also, as Apache Spark provides an interactive shell, learning Scala Shell will facilitate the process of learning Apache Spark down the road.

Checking Scala Installation Status in Your System

Chances are that no matter what operating system you are using—Linux, Windows, or Mac—Scala won't come preinstalled on it. Most operating

© Irfan Elahi 2019
I. Elahi, *Scala Programming for Big Data Analytics*,
https://doi.org/10.1007/978-1-4842-4810-2_2

systems come with the bare minimum software utilities and then developers install software, framework, and utilities on them to configure them as per their requirements. Each language has its own set of prerequisites and requirements for it to execute. We will initially focus on verifying whether Scala is installed on your system and, if not, how to set up the requirements.

It's not difficult to determine whether Scala is installed on your system. As I am currently using Microsoft Windows (specifically Windows 10, but it won't matter much in this context), the instructions are tailored to the Windows operating system. However, the process is similar regardless of the operating system and I strive to call out wherever appropriate when the process is different.

So, to quickly determine whether Scala is installed on your system, you open the command prompt from your Start menu (by opening the Start menu and either finding Command Prompt or typing cmd and pressing Enter). Once the command prompt opens, type scala and see what you get in the output, as illustrated in Figure 2-1.

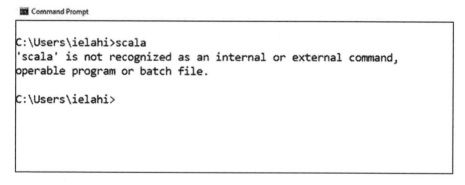

Figure 2-1. *Verifying Scala installation status on Windows*

If you get something similar to the screen shown in Figure 2-1, you can pretty much assume that Scala isn't installed on your system. Just a note: Scala could still be installed on your system, depending on the system's environment variables. We'll ignore that nuance and assume that Scala, as per this test, isn't installed on your system. So it's a clean slate to proceed to the next steps.

So what do you do next? Install Scala, of course.

Verifying Java Development Kit (JDK) Installation Status

There can be many ways you can go about installing Scala, but my intent here is to highlight the quickest and simplest way to do it so that you are up-to-speed in developing your first Scala program.

Before you jump right in to installing Scala on your system, knowing that you are quite tempted to get your hands dirty writing your first Scala program, you need to ensure that the prerequisites of Scala are met in your system.

Scala, as of this date, requires the Java Development Kit (JDK) to be installed on your system. To ensure you have the latest version, it's good practice to refer to the Scala and JDK compatibility documentation online (for instance, at `https://docs.scala-lang.org/overviews/jdk-compatibility/overview.html`).

For this chapter, I use JDK 8 because this is the recommended version as per Scala's download page. However, you can install any compatible version of JDK after researching what version you are trying to install and determining the recommended JDK for it. As Scala is a JVM language and whatever you write in Scala is converted to Java code, it's natural to have JDK/Java installed in your system.

To check if the JDK is installed on your system, open the command prompt and type:

```
java -version
```

The results are highlighted in Figure 2-2.

```
C:\Users\ielahi>java -version
java version "1.8.0_171"
Java(TM) SE Runtime Environment (build 1.8.0_171-b11)
Java HotSpot(TM) 64-Bit Server VM (build 25.171-b11, mixed mode)

C:\Users\ielahi>
```

Figure 2-2. *Verifying JDK installation status*

If you get output similar to what's shown in Figure 2-2, it means that Java is installed on your system. If you did not get this output, you have to install the JDK, as detailed in the next section.

What you basically did with the `java -version` command was check which version of Java you have. In simpler terms, when you install the JDK, it installs Java, among other components, for you. Note that your Java version might not exactly match the one used here. You might end up installing a newer version of the JDK, which is available at the time you are reading this book. However, as mentioned before, this should not be an issue as long as you are installing the latest available version of the JDK while factoring in what Scala supports by going through its requirements on its website.

Installing the Oracle JDK

If the JDK isn't installed in your system, you have to install it using the next steps.

The JDK is available by multiple vendors and you can choose from two major options: Oracle JDK and OpenJDK. For various reasons, I suggest you install Oracle JDK instead of OpenJDK. To elaborate that further, many Big Data vendors like Cloudera strictly recommend you opt for Oracle JDK as they don't support OpenJDK. So if you are on your way to pivot to Big Data analytics, it's good to stay aligned with what's recommended in this space.

The steps to installing Oracle JDK are pretty straightforward, on Windows at least. It's equivalent to installing any software, whereby you download the setup from a website and follow the installation wizard. Thus, go to Oracle's website. At the time of writing this book, the link to download the 1.8 version of JDK is:

```
http://www.oracle.com/technetwork/java/javase/downloads/jdk8-
downloads-2133151.html
```

As per your Windows operating system (i.e., x86 or x64), download the corresponding installer .exe file. Once it's downloaded, open the installer and follow along with the instructions from the installation wizard. This shouldn't be hard.

If you are using another operating system, use the appropriate installation instructions on Oracle's website.

Once the installation process is done, verify it by issuing the `java -version` command. If the JDK installed correctly, you will get a response similar to what's shown in Figure 2-2.

Installing Scala on Windows

Fortunately for Windows users, the process of installing Scala bears a strong resemblance to installing almost any other software. After you have ensured that you have the right version of Oracle JDK installed on Windows, you are in a good position to install Scala.

To install Scala on Windows, open your favorite web browser and go to the following URL:

https://www.scala-lang.org/download/)

Once you're there, you will be presented with many ways to install Scala. For instance, you can use the IDE (IntelliJ) or SBT. However, at this stage, let's skip those options. Search for **Other ways to install Scala** on that web page and click the **Download the Scala Binaries for Windows** option, as shown in the Figure 2-3.

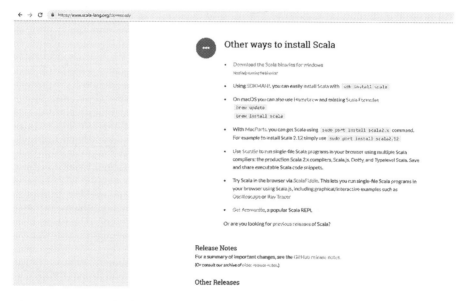

Figure 2-3. *Scala download option available on the Scala website*

When you install Scala this way, you will be able to use it from the Windows command prompt (which, if you recall, complained previously when you tried typing Scala there) and you will be able to launch the Scala shell called REPL (which stands for Read-Eval-Print-Loop - it's an interactiv shell which we will be using a lot in this book), among other components. So it's a quick way to get up and running with Scala without using the other installation options such as SBT and IntelliJ IDEA.

After you have downloaded the Scala installer from the website, open the downloaded setup/installer file. It will greet you with a screen similar to what's shown in Figure 2-4.

Figure 2-4. *Scala installation wizard*

You may have guessed the next steps already? But for your reference, they are presented as follows:

1. Click Next.

2. Read the end user license agreement and check **I Accept the Terms in the License Agreement**. Then click Next.

3. Select a location on your system where you want Scala to be installed and then click Install.

After a while, Scala should be installed on your system successfully.

Verifying Scala Installation Status

Upon successful installation, of any software, it helps to verify it's installation status. Thus, to check whether Scala is installed successfully in your system, you can proceed as follows.

Open the Command Prompt again and type scala (it's case sensitive, so make sure you type scala in lowercase). You should get a screen similar to what's shown in Figure 2-5.

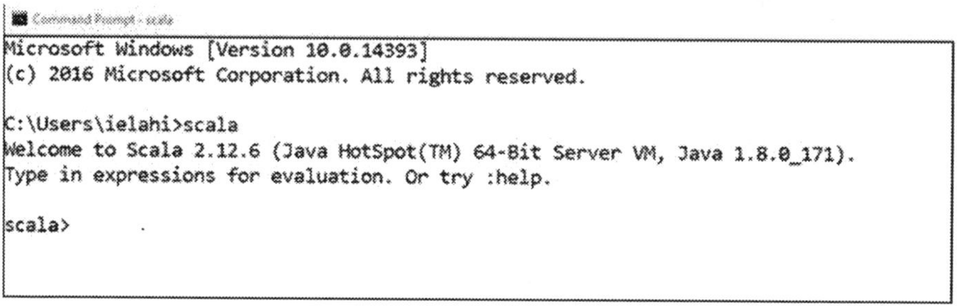

```
Command Prompt - scala
Microsoft Windows [Version 10.0.14393]
(c) 2016 Microsoft Corporation. All rights reserved.

C:\Users\ielahi>scala
Welcome to Scala 2.12.6 (Java HotSpot(TM) 64-Bit Server VM, Java 1.8.0_171).
Type in expressions for evaluation. Or try :help.

scala>
```

Figure 2-5. *Verifying Scala's successful installation on Windows*

If you still get an error when using the scala command in your command prompt, you can add scala bin folder (which is located in the folder where you installed Scala) to your path's environment variable in your system.

Most likely, you'll see that Scala is working fine and it has launched a Scala shell (REPL) for you. It also provides additional information, such as which version is installed and which JDK it's using. The JDK version will be the same one you installed previously.

Installing Scala on Linux

Given that Linux is being heavily used in professional environments to power a diverse variety of applications (including Apache Spark), this section covers installing Scala on Linux machines. There are a number of Linux distributions and it will not be possible to cover installation instructions for all of them, but I demonstrate how to install it on one of the distros (Ubuntu 18.04.2 LTS); you can follow the same concepts to install Scala on other distros.

For this installation, I assume that you are using the Linux operating system. If not, you can use a Virtualization solution (e.g., use Ubuntu's image in an Oracle Virtual Box) to run Linux on your Windows operating system. Provided that you have a Linux environment handy, the steps to install Scala are as follows.

Just like with Windows, you need to have the JDK installed on your Linux operating system. The steps to install the JDK in Ubuntu are as follows.

Open the Ubuntu shell and issue the following commands to update all existing packages on your Ubuntu system.

```
sudo su
```

```
apt-get update
```

It's always a recommended practice to do so. Also, it is helpful if you change your user to root so that you can issue these commands without running into privileges issues (i.e., your user may not have the privilege/authorization to run such system commands). You can run each of the following commands with sudo as well.

Figure 2-6 highlights the example output after running the apt-get update command.

```
root@ievm2:~# apt-get update
Hit:1 http://azure.archive.ubuntu.com/ubuntu bionic InRelease
Get:2 http://azure.archive.ubuntu.com/ubuntu bionic-updates InRelease [88.7 kB]
Get:3 http://azure.archive.ubuntu.com/ubuntu bionic-backports InRelease [74.6 kB]
Get:4 http://azure.archive.ubuntu.com/ubuntu bionic/multiverse Sources [181 kB]
Get:5 http://security.ubuntu.com/ubuntu bionic-security InRelease [88.7 kB]
Get:6 http://azure.archive.ubuntu.com/ubuntu bionic/restricted Sources [5324 B]
Get:7 http://azure.archive.ubuntu.com/ubuntu bionic/main Sources [829 kB]
Get:8 http://azure.archive.ubuntu.com/ubuntu bionic/universe Sources [9051 kB]
Get:9 http://azure.archive.ubuntu.com/ubuntu bionic-updates/main Sources [252 kB]
Get:10 http://azure.archive.ubuntu.com/ubuntu bionic-updates/multiverse Sources [4192 B]
Get:11 http://azure.archive.ubuntu.com/ubuntu bionic-updates/restricted Sources [2068 B]
Get:12 http://azure.archive.ubuntu.com/ubuntu bionic-updates/universe Sources [135 kB]
Get:13 http://azure.archive.ubuntu.com/ubuntu bionic-updates/main amd64 Packages [545 kB]
Get:14 http://azure.archive.ubuntu.com/ubuntu bionic-updates/main Translation-en [203 kB]
Get:15 http://azure.archive.ubuntu.com/ubuntu bionic-updates/restricted amd64 Packages [6996 B]
Get:16 http://azure.archive.ubuntu.com/ubuntu bionic-updates/universe amd64 Packages [740 kB]
Get:17 http://azure.archive.ubuntu.com/ubuntu bionic-updates/universe Translation-en [191 kB]
Get:18 http://security.ubuntu.com/ubuntu bionic-security/universe Sources [35.1 kB]
Get:19 http://azure.archive.ubuntu.com/ubuntu bionic-updates/multiverse amd64 Packages [6388 B]
Get:20 http://azure.archive.ubuntu.com/ubuntu bionic-backports/universe Sources [2068 B]
Get:21 http://security.ubuntu.com/ubuntu bionic-security/main Sources [76.4 kB]
Get:22 http://security.ubuntu.com/ubuntu bionic-security/restricted Sources [1504 B]
Get:23 http://security.ubuntu.com/ubuntu bionic-security/multiverse Sources [2308 B]
Get:24 http://security.ubuntu.com/ubuntu bionic-security/restricted amd64 Packages [4296 B]
Get:25 http://security.ubuntu.com/ubuntu bionic-security/restricted Translation-en [2192 B]
Get:26 http://security.ubuntu.com/ubuntu bionic-security/universe amd64 Packages [126 kB]
Get:27 http://security.ubuntu.com/ubuntu bionic-security/multiverse amd64 Packages [3744 B]
Get:28 http://security.ubuntu.com/ubuntu bionic-security/multiverse Translation-en [1952 B]
Fetched 12.7 MB in 5s (2428 kB/s)
Reading package lists... Done
root@ievm2:~# apt-get upgrade
Reading package lists... Done
Building dependency tree
Reading state information... Done
Calculating upgrade... Done
```

Figure 2-6. *Updating Ubuntu packages*

Once the process is complete, issue the following command to add the Oracle JDK to your list of repositories.

```
add-apt-repository ppa:webupd8team/java
```

This command is required for Ubuntu's package management system to find Java from this repository when you issue installation commands. After you issue this command, the process may prompt you to accept the license agreement. Accept the agreement after reading it; the installation will resume.

After the installation is complete, issue the following command to start the installation of Oracle JDK:

```
apt install oracle-java8-installer
```

Once this is done, verify the installation status of the Oracle JDK by issuing the `java -version` command.

It should show output similar to Figure 2-7.

```
root@ievm2:~$ java -version
java version "1.8.0_201"
Java(TM) SE Runtime Environment (build 1.8.0_201-b09)
Java HotSpot(TM) 64-Bit Server VM (build 25.201-b09, mixed mode)
```

***Figure 2-7.** Java version in Ubuntu indicating successful installation*

Lastly, to install Scala, issue the following command:

```
apt-get install scala
```

Figure 2-8 shows the sample output of this command.

```
root@ievm2:~$ apt-get install scala
Reading package lists... Done
Building dependency tree
Reading state information... Done
The following additional packages will be installed:
  libhawtjni-runtime-java libjansi-java libjansi-native-java libjline2-java scala-library scala-parser-combinators scala-xml
Suggested packages:
  scala-doc
The following NEW packages will be installed:
  libhawtjni-runtime-java libjansi-java libjansi-native-java libjline2-java scala scala-library scala-parser-combinators scala-xml
0 upgraded, 8 newly installed, 0 to remove and 0 not upgraded.
Need to get 25.1 MB of archives.
After this operation, 28.7 MB of additional disk space will be used.
Do you want to continue? [Y/n] Y
Get:1 http://azure.archive.ubuntu.com/ubuntu bionic/universe amd64 libhawtjni-runtime-java all 1.15-2 [27.1 kB]
Get:2 http://azure.archive.ubuntu.com/ubuntu bionic/universe amd64 libjansi-native-java all 1.7-1 [19.4 kB]
Get:3 http://azure.archive.ubuntu.com/ubuntu bionic/universe amd64 libjansi-java all 1.16-1 [36.2 kB]
Get:4 http://azure.archive.ubuntu.com/ubuntu bionic/universe amd64 libjline2-java all 2.14.6-1 [150 kB]
Get:5 http://azure.archive.ubuntu.com/ubuntu bionic/universe amd64 scala-library all 2.11.12-2 [9596 kB]
Get:6 http://azure.archive.ubuntu.com/ubuntu bionic/universe amd64 scala-parser-combinators all 1.0.3-3 [355 kB]
Get:7 http://azure.archive.ubuntu.com/ubuntu bionic/universe amd64 scala-xml all 1.0.3-3 [601 kB]
Get:8 http://azure.archive.ubuntu.com/ubuntu bionic/universe amd64 scala all 2.11.12-2 [14.3 MB]
Fetched 25.1 MB in 2s (13.1 MB/s)
Selecting previously unselected package libhawtjni-runtime-java.
(Reading database ... 56208 files and directories currently installed.)
Preparing to unpack .../0-libhawtjni-runtime-java_1.15-2_all.deb ...
Unpacking libhawtjni-runtime-java (1.15-2) ...
Selecting previously unselected package libjansi-native-java.
Preparing to unpack .../1-libjansi-native-java_1.7-1_all.deb ...
Unpacking libjansi-native-java (1.7-1) ...
Selecting previously unselected package libjansi-java.
Preparing to unpack .../2-libjansi-java_1.16-1_all.deb ...
Unpacking libjansi-java (1.16-1) ...
Selecting previously unselected package libjline2-java.
Preparing to unpack .../3-libjline2-java_2.14.6-1_all.deb ...
Unpacking libjline2-java (2.14.6-1) ...
Selecting previously unselected package scala-library.
Preparing to unpack .../4-scala-library_2.11.12-2_all.deb ...
Unpacking scala-library (2.11.12-2) ...
Selecting previously unselected package scala-parser-combinators.
Preparing to unpack .../5-scala-parser-combinators_1.0.3-3_all.deb ...
Unpacking scala-parser-combinators (1.0.3-3) ...
Selecting previously unselected package scala-xml.
Preparing to unpack .../6-scala-xml_1.0.3-3_all.deb ...
Unpacking scala-xml (1.0.3-3) ...
Selecting previously unselected package scala.
Preparing to unpack .../7-scala_2.11.12-2_all.deb ...
Unpacking scala (2.11.12-2) ...
Setting up scala-parser-combinators (1.0.3-3) ...
Setting up libhawtjni-runtime-java (1.15-2) ...
```

***Figure 2-8.** Installing Scala on Ubuntu*

After installation, verify Scala's installation status by issuing `scala` command in the shell. It should display Scala REPL. Figure 2-9 highlights that result.

```
root@ievm2:~# scala
Welcome to Scala 2.11.12 (Java HotSpot(TM) 64-Bit Server VM, Java 1.8.0_201).
Type in expressions for evaluation. Or try :help.

scala>
```

Figure 2-9. *Scala REPL in Ubuntu*

Congratulations. At this stage, you have successfully installed Scala on Windows or on Linux. You have configured a development environment on your system, which you can use to follow along with this book.

EXERCISES

- Figure out how to install Scala on Linux in an offline setting (e.g., on a system that's not connected to the Internet). You will encounter such scenarios extensively in professional environments due to security controls.

- Type `scala -help` at the command prompt and familiarize yourself with as many options available as possible.

- Try installing multiple versions of Scala on your system and investigate if there are any challenges.

- Research the latest versions of Scala available on the website and make a habit of going through the release notes to understand the latest enhancements in the new releases.

CHAPTER 3

Using the Scala Shell

In the developer communities in nearly all parts of the world, productivity is deemed one of the critical key performance indicators (KPIs). If the experience of using a tool or language enhances a developer's productivity, this is considered as a strong plus for that tool or language. There can be many language characteristics that increase a developer's productivity, and one of the reasons Scala is loved by many is because of its REPL/shell feature.

If you have used Java before, you may know that in order to write and execute a program, even one as simple as printing Hello World, you have to follow all the involved steps. Those steps include, but are not limited to, creating a .java file (ensuring that the filename is equal to the class within that file). Then you compose a class in that file with a `main` method (if you want to make it executable) and then compile it (resulting in Java bytecode) and run it. If you want to try a new expression in Java, you have to do all of these steps again and again. By using IDEs like Eclipse, the process is quick but the steps are the same.

In Scala, the situation is rather different. By virtue of the Scala shell, you can type an expression and get the result right away, thus eliminating many of the steps mentioned for Java (just an FYI that Java 9 launched the JShell for this same purpose). A similar experience exists in Python as well. However, note that this Scala use case is well suited for exploratory and experimentation purposes or even for Data Scientists or Data Engineers who want to quickly prototype something. For production deployments, you follow similar steps as needed in Java (although there is high degree of flexibility that Scala furnishes, as we'll see).

© Irfan Elahi 2019
I. Elahi, *Scala Programming for Big Data Analytics*,
https://doi.org/10.1007/978-1-4842-4810-2_3

In the previous chapter, we walked through the process of installing Scala on a Windows and Linux system. The chapter also covered how to verify Scala installation and how to launch the Scala shell/REPL. (I use these terms interchangeably.) With the Scala shell launched, you are all set to do programming in a highly interactive environment, which will foster your learning process. The Scala shell, once launched, will be ready to accept your commands or Scala expressions, execute them, and display the results immediately, without any delay. To emphasize further the ubiquity of the Scala shell, I use it when I want to try specific expressions on an ad hoc basis prior to productionizing my codebase. The Scala shell is my go-to tool for this purpose, as it significantly increases my productivity. When you explore Apache Spark, you will find that it also comes with a shell (for both Scala and Python). Spark's shell for Scala is the same as the Scala shell that you'll use in this chapter.

Let's get familiar with this tool before we move on to the next stages of our learning journey, shall we? As a first step, launch the Scala shell on your system. (It should be clear by now that you can launch Scala by opening the Windows Command Prompt (or Linux shell) and typing `scala`.)

Getting Help from the Scala Shell

One of the first things that I do when using a new tool is determine which options or commands I have available at my disposal. Every command line interface (CLI) tool, i.e., a tool that you use via the Command Prompt, comes with a number of options that you can use to your advantage. Having a certain degree of familiarity with those options always helps.

Once you have launched the Scala shell, type `:help` in the shell. Make sure not to miss the colon (the `:`) before the **help** keyword. There should be no spaces between `:` and `help`, as shown in Figure 3-1.

```
scala> :help
All commands can be abbreviated, e.g., :he instead of :help.
:completions <string>     output completions for the given string
:edit <id>|<line>         edit history
:help [command]           print this summary or command-specific help
:history [num]            show the history (optional num is commands to show)
:h? <string>              search the history
:imports [name name ...] show import history, identifying sources of names
:implicits [-v]           show the implicits in scope
:javap <path|class>       disassemble a file or class name
:line <id>|<line>         place line(s) at the end of history
:load <path>              interpret lines in a file
:paste [-raw] [path]      enter paste mode or paste a file
:power                    enable power user mode
:quit                     exit the interpreter
:replay [options]         reset the repl and replay all previous commands
:require <path>           add a jar to the classpath
:reset [options]          reset the repl to its initial state, forgetting all session entries
:save <path>              save replayable session to a file
:sh <command line>        run a shell command (result is implicitly => List[String])
:settings <options>       update compiler options, if possible; see reset
:silent                   disable/enable automatic printing of results
:type [-v] <expr>         display the type of an expression without evaluating it
:kind [-v] <type>         display the kind of a type. see also :help kind
:warnings                 show the suppressed warnings from the most recent line which had any
```

Figure 3-1. *Output of :help in Scala REPL*

It will display all the possible commands, apart from Scala expressions, that you can type here. I won't cover all of them but will highlight just a couple of them in this chapter (and the book). You can research others on your own, as needed.

Hello World in Scala REPL

Now let's do what every new programmer does when they start learning programming, i.e., let's print hello world! Let's do it in the Scala shell and see how difficult or easy it is.

To print Hello World, just type "hello world" in the shell, as shown in Figure 3-2.

```
scala> "hello world"
res3: String = hello world
```

Figure 3-2. *Printing hello world in Scala REPL*

You will get output as shown in Figure 3-2. There you have it. You have displayed hello world programmatically using Scala in your environment. Technically you didn't "print" it like you normally do with print function, but for now, it's more than enough.

Understanding Hello World in Scala REPL Step-by-Step

Now what just happened in the previous step? Let's take a look at the process in a step-by-step manner. Understanding this will form a strong foundation for your understanding of Scala and its shell:

- You actually typed a *string* in the Scala shell. String is one of the data types available in Scala and you'll use a combination of different data types in your programs to achieve the task that you want to perform. You use the String data type to represent alphanumeric characters like "hello world". Strings in Scala are represented by double quotes, i.e. "". It's different from Python, where Python accepts both single and double quotes for strings. In Scala, strings must be enclosed in double quotes (or triple quotes, as we'll see in coming chapters).

- Whatever you type in a double quote, it becomes a String data type. In this case when you typed

  ```
  "hello world"
  ```

  And pressed Enter, Scala displayed this output on the screen:

  ```
  res3: String = hello world
  ```

- What does that output mean that was displayed as a result of you typing "hello world"? It means that the Scala shell created a variable for you named res3. Every time you type and enter anything in the Scala shell, it is assigned to a variable, which generally starts with resN (where res stands for result and N can be any number depending on how many expressions you have entered in that shell session). What is a variable, you may inquire? Although we haven't covered that concept in detail, you can imagine it to be a holder or container of data, i.e., something that has a name and can store data in it.

- Another important piece of information that you can gather from here is that res3, the variable that was created, is of type String. Scala is a strongly typed language, as explained in the first chapter, and each variable has a type associated with it. In this case, the type of the variable, obviously, is String, and this fact is confirmed by the Scala shell. This particular feature of the Scala shell is particularly useful as having an understanding of data type is crucial for you to write logically correct programs. Computer programs are all about operating on data of varying types, so knowing what type of data you are manipulating will help you choose the correct method and will avoid many errors.

Given that the Scala shell created a variable for you when you simply typed "hello world", you can access and use that variable if you want to. Just type res3 in the shell and see what you get for output (it will be similar to Figure 3-3).

```
scala> res3
res4: String = hello world

scala>
```

Figure 3-3. *Displaying the res3 variable that Scala REPL created*

Sure enough, you'll get hello world in the output along with other information. This is a great way to understand the type and value of an expression that is generated when you type Scala expressions.

In this case, we simply typed "hello world" but you can try any valid Scala expression to see what value and type is returned. Just type 1+10 in the shell, as highlighted in Figure 3-4.

```
scala> 1+10
res5: Int = 11

scala>
```

Figure 3-4. *Using a mathematical expression in Scala REPL*

This time, it shows that it created a variable called res5. It's of type INT (which is one of the numerical types in Scala used to deal with numbers), and its value is 11. This time you created a variable of type Integer (or Int to be specific; I use Integer and Int interchangeably) instead of String.

I must say that you will rarely use these variables (like res4 and res5) that the Scala shell creates, but what's important is to see the result of expressions and the types. This notion will become more important if you do advanced programming in Scala, as discussed in the next section.

Using Scala REPL's Data Type Highlighting Feature

Let's look at a slightly complex yet real-world example to show how the feature of the Scala shell that highlights data type can be helpful in many settings.

A while ago, I was working on a Scala program where I had to retrieve database name from a variable that included database name and table name with the following naming convention:

```
database_name.table_name
```

I wanted to retrieve just database name. For those who know programming, you may have guessed how to do this. One of the ways to do this is to use the split function, which is available in the String type. If you have a string like "Irfan_Pakistan" for example and if you use the "Irfan_Pakistan".split("_") function, it will split at the underscore _ character. Once it's split, you can get the element of your choice (either Irfan or Pakistan).

Now when you use "Irfan_Pakistan".split("_"), the result of this expression isn't a string. It's something else. Figure 3-5 shows how you can use the Scala shell to see the type of the result of using the split function on String.

```
scala> "Irfan_Pakistan".split("_")
res11: Array[String] = Array(Irfan, Pakistan)

scala>
```

Figure 3-5. *Highlighting how to use the split method on a String data type*

It indicates that when we use the split function on a string, we'll get an Array in return.

Note   An Array is another type in Scala that's used when you want a variable to store a collection of values. We'll learn what collections are in upcoming chapters, but for now, just keep in mind that they are different from variables of other data types like Integer, which can hold only one value.

Knowing that the expression gives me Array, I can then use it to get my desired values out of it (by using indexing—think of this like retrieving a particular element from an Array collection), as shown in Figure 3-6.

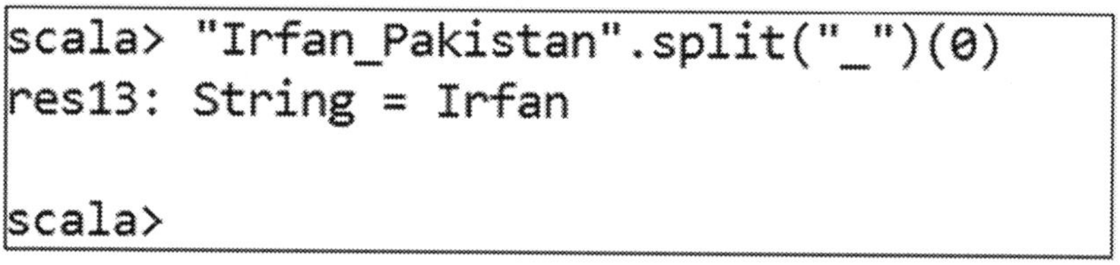

```
scala> "Irfan_Pakistan".split("_")(0)
res13: String = Irfan

scala>
```

Figure 3-6. *Indexing the array that gets returned via the split method*

Note that we used (0) in the expression in Figure 3-6, which indicates that we are getting the first element of the collection (Array in this case). For now, it's enough to know that Scala uses zero-based indexing (which in simpler terms means that the first element of a Scala collection like Array is element 0, the second element is element 1, and so on).

If you didn't know already, now you know the implications of using split, i.e., it returns an Array.

Now with this contextual knowledge, we can attack the problem at hand. If we use

```
"sample_db.my_table".split(".")
```

It should give us an array like before, right? Well, go on and give it a try and see what happens. If you do, you will see that you will, interestingly, get an exception. This means that what you are trying to do is wrong. Congrats, you've lived long enough to see your first Scala exception, as shown in Figure 3-7.

```
scala> "sample_db.my_table".split(".")(0)
java.lang.ArrayIndexOutOfBoundsException: 0
  ... 28 elided

scala>
```

Figure 3-7. *An example of an ArrayIndexOutOfBounds exception*

The point is, you expected the split function to work the same way but it didn't and you checked it with the Scala shell instantly. (A more appropriate way of doing this would be to write unit test cases, which is something that you should look into once you become slightly more well-versed with Scala.) What happened is that, in this particular instance, you need to do something different to achieve the desired result, as shown in Figure 3-8.

```
scala> "sample_db.my_table".split("\\.")(0)
res15: String = sample_db

scala>
```

Figure 3-8. *The correct way to split a string on a . character*

In the example shown in Figure 3-8, I used similar constructs as before with one variation that I used ("\\0") while splitting. Using . in this context without \\ was interpreted differently by Scala. It indicated that you wanted to use regular expressions (which are used for matching patterns in strings). Rather, the intent was to just split on the **.** character and not involve regular expressions in any way. That's why we "escaped" the **.** character by using the \\ to indicate our intentions to Scala.

The key takeaway is that using the Scala shell, you can see the type of the result of the expression and, if possible, detect errors. Which, trust me, is really handy!

Paste Mode in Scala REPL

Now back to the rhythm: You can use this shell to type any Scala expression that you want and see the results instantaneously. It works great for single line expressions, but you can run into issues when you are trying to paste in multiple lines of code. For instance, say you wrote a few lines of Scala code in a text editor (e.g., notepad or Sublime) and you want to execute them in the Scala shell in one go. You can just paste them in the shell, but there is a better way to do this. That's where the paste mode comes into play.

To use paste mode, type `:paste` in the shell. This brings you into paste mode, where you can paste or type multiple expressions and execute all of them at once by pressing Ctrl+D, as shown in Figure 3-9.

```
scala> :paste
// Entering paste mode (ctrl-D to finish)

val x = "irfan"
println(x)
println(x.toUpperCase + "ELAHI")

// Exiting paste mode, now interpreting.

irfan
IRFANELAHI
x: String = irfan

scala>
```

Figure 3-9. *Paste mode in Scala REPL*

In Figure 3-9, I initialized a variable named x in Scala. val before x stands for value and it means that the variable that you are creating is immutable, i.e., its value can't be changed or you can't assign a value back to this variable. But at a higher level, it's how you generally initialize a variable in Scala. You can initialize using var as well, which creates a mutable variable. We'll look at these concepts in more detail in subsequent chapters.

Then I used two print statements to print the value of the variable. In the second print statement, I used the .toUpperCase function of String (just like I used split function before) to convert the string to uppercase; I also used the + sign to *concatenate* (i.e., join together) the two strings.

If you are working with classes and traits in Scala, paste mode is very handy. Similarly, for control statements (like if-else conditionals and loops), it proves to be useful.

Retrieving History in Scala REPL

Whatever you type in the Scala shell is stored and you can retrieve the history as well. The experience is similar to the commands history in the Linux shell. It's helpful when you want to see which commands/expressions you executed previously and reuse them by copying and pasting them back into the shell.

To see a history of expressions that you typed in the Scala shell, type :history in the shell, as shown in Figure 3-10.

```
scala> :history
884    :paste
885    "this is first string"
886    1+2
887    "first_second".split("_")
888    :paste
889    val x="irfan"
890    print(x)
891    print(x.toUppercase)
892    x.toUpperCase
893    x
894    "h".toUpperCase
895    :paste
896    val x = "irfan"
897    print(x)
898    print(x.toUpperCase)
899    :paste
900    val x = "irfan"
901    println(x)
902    println(x.toUpperCase + "ELAHI")
903    :history
```

Figure 3-10. *Getting a history of expressions in Scala REPL*

You can also type a number after :history to inform the Scala shell that you intend to get that many items from history, as illustrated in Figure 3-11.

```
scala> :history
  1  case class EmployeeData(designation:String, company:String)
  2  val employeeData = List("engineer,facebook","manager,facebook","associate,facebook")
  3  employeeData.map(x=>x.split(",")).map(x=>EmployeeData(x(0),x(1)))
  4  scala> val employeeData = List("engineer,facebook","manager,facebook","associate,facebook")
  5  employeeData: List[String] = List(engineer,facebook, manager,facebook, associate,facebook)
  6  val employeeList = employeeData.map(x=>x.split(",")).map(x=>EmployeeData(x(0),x(1)))
  7  case class EmployeeData(designation:String, company:String)
  8  val employeeList = employeeData.map(x=>x.split(",")).map(x=>EmployeeData(x(0),x(1)))
  9   val employeeData = List("engineer,facebook","manager,facebook","associate,facebook")
 10  val employeeList = employeeData.map(x=>x.split(",")).map(x=>EmployeeData(x(0),x(1)))
 11  employeeList(0)
 12  employeeList(0).designation
 13  employeeList.map(x=>x.designation)
 14  :help
 15  val myStringVariable = "irfan"
 16  :historuy
 17  :history
scala>
```

Figure 3-11. *Using the history feature with a number to specify the number of items you want to retrieve*

If you type :history 200, it will display the past 200 expressions. It's a pretty handy feature to see what you typed before.

The Auto-Completion Feature of Scala REPL

When you search using major search engines like Google, the search engine keeps on suggesting or auto-filling your intended search query. This streamlines the user experience significantly. Similarly, you saw in the previous examples that the String data type had some functions associated with it (we used .toUpperCase and the .split method to do different types of tasks).

How can we determine which methods are available to different data types/objects in Scala? One option is to keep the Scala API documentation handy, but in many instances, you'll quickly want to know what's available in a particular object. To facilitate that, you can employ the Scala shell's auto-completion feature.

Say that you created a variable of type `integer` in a Scala REPL session as follows:

```
val myVariable = 20
```

Then, when you type some characters of your defined variable like `myVar` and press Tab, Scala REPL will auto-complete it for you. It can add a lot to your productivity and can save you some typing (and prevent errors).

Another interesting and helpful extension of this feature is that it allows you to see the functions and fields available in your object. I know you may not have a thorough understanding of object oriented programming and the terms like functions, fields, and object may not make such sense to you right now. But at a higher level, whenever you create a variable in Scala REPL, for example,

```
val myStringVariable = "Scala"
```

it creates a variable of a specific type (`String` in this case). Thus, depending on the nature of the variable created, e.g. `Integer`, `String`, etc., there are a number of methods (similar to functions, which you will study later in the book, but think of them as modules of code that do specific tasks) and fields (also known as attributes, and they represent properties of that object) that come with each type.

If you want to see which functions and fields are available for the `String` variable that you just created, type the variable, then type . and then press Tab. It will display a list of what's available for the `String` type, as shown here:

```
scala> myStringVariable.
!=   compareTo          genericBuilder  matches             runWith
toBuffer
##   compareToIgnoreCase getBytes        max                 sameElements
toByte
*    compose            getChars         maxBy               scan
toCharArray
+    concat             getClass         min                 scanLeft
toDouble
++   contains           groupBy          minBy               scanRight
toFloat
++:  containsSlice      grouped          mkString            segmentLength
toIndexedSeq
+:   contentEquals      hasDefiniteSize ne                   self
toInt
->   copyToArray        hashCode         nonEmpty            seq
toIterable
/:   copyToBuffer       head             notify              size
toIterator
:+   corresponds        headOption       notifyAll           slice
toList
:\   count              indexOf          offsetByCodePoints  sliding
toLong
<    diff               indexOfSlice     orElse              sortBy
toLowerCase
<=   distinct           indexWhere       padTo               sortWith
toMap
==   drop               indices          par                 sorted
toSeq
>    dropRight          init             partition           span
toSet
>=   dropWhile          inits            patch               split
toShort
```

41

addString	endsWith	intern	permutations
splitAt	toStream		
aggregate	ensuring	intersect	prefixLength
startsWith	toString		
andThen	eq	isDefinedAt	product
stringPrefix	toTraversable		
apply	equals	isEmpty	r
stripLineEnd	toUpperCase		
applyOrElse	equalsIgnoreCase	isInstanceOf	reduce
stripMargin	toVector		
asInstanceOf	exists	isTraversableAgain	reduceLeft
stripPrefix	transpose		
canEqual	filter	iterator	reduceLeftOption
stripSuffix	trim		
capitalize	filterNot	last	reduceOption
subSequence	union		
charAt	find	lastIndexOf	reduceRight
substring	unzip		
chars	flatMap	lastIndexOfSlice	reduceRightOption
sum	unzip3		
codePointAt	flatten	lastIndexWhere	regionMatches
synchronized	updated		
codePointBefore	fold	lastOption	replace
tail	view		
codePointCount	foldLeft	length	replaceAll
tails	wait		
codePoints	foldRight	lengthCompare	replaceAllLiterally
take	withFilter		
collect	forall	lift	replaceFirst
takeRight	zip		
collectFirst	foreach	lines	repr
takeWhile	zipAll		
combinations	format	linesIterator	reverse
to	zipWithIndex		

companion	formatLocal	linesWithSeparators	reverseIterator
toArray	?		
compare	formatted	map	reverseMap
toBoolean			

As you can see, there is a thorough list showing what you can access/invoke on the object. For example, to determine the length of your string variable's value, you can call:

```
myStringVariable.length
```

Similarly, to find the location/index of a specific character in your string (e.g., to find where the character "c" appears in your string variable value "Scala"), you can use another method that comes with the String type:

```
myStringVariable.indexOf("c")
```

In the output, you will get 1 as the result. 1, in this context, means that it exists in the second position from the left. Remember, counting positions (called indexing) starts at zero in Scala.

Thus, using this feature of Scala REPL, you can quickly determine which fields or methods are available.

Exiting from Scala REPL

On a lighter note, unlike many text editors like Vim, it's quite easy to exit from the Scala shell.

To quit the Scala shell/REPL, you simply type:

```
:quit
```

Technically, at the backend, your OS launches a Java Virtual Machine process that runs the Scala shell and, as you exit the session, that JVM process is killed as well.

Well, that's enough for Scala REPL. Let's move on to the next chapter and learn about some fundamentals of the Scala language in a proper and methodical manner, shall we?

EXERCISES

- Create a variable of type Int, assign a value, and then see what methods are available for that type. Repeat the same process for the String variable.

- Explore other options that are available in Scala REPL and research how you can use them (e.g., :sh, :save, :load, etc.).

- Research the spark-shell and see if the same commands and features are available there as well.

- Try increasing the memory used by the Scala shell.

CHAPTER 4

Variables

When you write programs, you use variables and you use a lot of them. The notion of variables was introduced briefly in the previous chapter. You use them to refer to different objects that you create. For example, if you are writing a program to store the result of the mathematical expression 10+5, you will typically store it in a container/placeholder so that you can reuse it with further operations. For instance, if you created a variable that stored an integer value, you could further use it for numerical operations like addition or subtraction. In the context of programming, these placeholders/containers are called *variables*, and that's what we will thoroughly explore in this chapter.

Getting Started with Variables in Scala

In the backend, depending on the language you are using and the type of variable that you are using, each variable creation results in memory allocation. So typically, in Scala, you will write something like this:

```
val sumResult = 10+5
```

Let's look at what's happening in this statement. First, refer to Figure 4-1.

© Irfan Elahi 2019
I. Elahi, *Scala Programming for Big Data Analytics*,
https://doi.org/10.1007/978-1-4842-4810-2_4

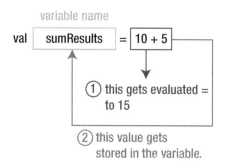

Figure 4-1. *Understanding variables and value assignments in Scala*

Here is a description of this statement:

- The right side of the equation, i.e. the elements on the right side of the equals sign, are evaluated or executed. This generates a result, which in this case is 15.

- On the left side of the equation, we define a variable with the name sumResult. sumResult is the name of the variable and you will refer to this variable with this name.

- The result of the right side of the equation (15) will be stored in the variable called sumResult. Technically, under the hood, it will store a reference to the location in memory (specifically, in the heap memory space of JVM), where 15 gets stored and its reference (address) is assigned to the variable name sumResult. However, for now, you can assume that the value 15 is stored in sumResult.

Immutability of Variables in Scala

Chances are that if you have worked in other programming languages like Python or Java, you have some concept and understanding of variables. If not, generally, it's assumed that values in variables can vary. Also, the literal meaning of a variable is something that can vary/change. This is a common practice in all programming languages—you create variables and then you change their values whenever it's appropriate. For example in Python, you can do something like this:

```
# example of using variables in Python:
sumResult = 10+5
sumResult = 9+100
```

Initially, the value of the sumResult variable was 15 and then, in the next expression, we changed it to something else (109). Convenient isn't it? For a string, in Python, you do it like this:

```
# another example of using variables in Python:
myName = "Irfan"
myName = "Irfan" + "Elahi"
```

The reason I gave quick examples from the Python language is to emphasize the following fact: *Variables are used in almost all programming languages and it's considered perfectly normal in many languages (even in Scala in certain scenarios) to change their values*. This characteristic is called *mutability*, i.e., the ability to mutate or change.

But overall, as you will discover, Scala emphasizes the *immutability* of variables compared to other languages, i.e., the values in variables should not change. This philosophy is deeply rooted in the fact that Scala is a functional programming language, so mutability is considered as a bad practice for certain reasons (which will be explained later in the book).

47

Defining Variables (Mutable and Immutable) in Scala

Now with this context in place, there are two ways you can create variables in Scala:

- Using the val keyword

- Using the var keyword

Simply speaking, if you create a variable with the val keyword, you can't change its value. In other words, it's immutable and can't be updated. Whereas if you create a variable with the var keyword, you can change its value whenever you want (it's mutable).

Let's play with this concept a bit more to develop an in-depth understanding. Open Scala REPL and create a variable named country preceded with the val keyword, as follows:

```
val country = "Pakistan"
```

Then try reassigning a different value to this variable, e.g. Australia, and see what you get. Do this:

```
country = "Australia"
```

As soon as you attempt this, Scala will respond with loud complaints that it's not possible. Specifically, it will report to you:

```
error: re-assignment to val
```

That is, in Scala, it's not possible to assign a value to an already initialized val variable.

Now, do these same steps but this time, use var instead of val and see if it works. Here's a hint: It should. When you use the var keyword, you are creating a mutable variable.

Take a look at the sample screenshot in Figure 4-2.

```
scala> var country = "Pakistan"
country: String = Pakistan

scala> country
res5: String = Pakistan

scala> country = "Australia"
country: String = Australia

scala> country
res6: String = Australia

scala>
```

Figure 4-2. *Mutability in variables created using the var keyword*

In this step, by creating a mutable variable, we could change the value of the country variable to Australia from Pakistan.

Why Is Immutability Emphasized in Scala?

So why bother having val or immutable variables in Scala? This relates to Scala's inclination toward functional programming constructs, where it frowns on the idea of impurity. In the world of functional programming, there is a dire threat that it addresses: side effects. By side effects, it means that if you define a variable and if the value of that variable is changed by any function somewhere, this is a side effect.

When there is possibility that any function can change the value of any variable at any time whenever desired, this leads to another severe consequence: during code maintenance it becomes significantly difficult to track what changed the variable state and when and why.

That's why, in functional programming, you will hear the term *pure functions* a lot, which represents the same idea. A function is pure if it doesn't create any side effects, i.e., it doesn't change the state of any variable outside of its scope.

Also, one of the design goals of Scala was to make it highly concurrent. Using mutable variables in programs where multiple threads/processes are trying to interact with them can lead to inconsistent states. Thus, it's considered best practice to use immutable variables in Scala.

So, Scala practitioners recommend using `val` wherever possible. Or, in fact, all of the time. In very specific scenarios where you can't get away with using a mutable variable, only then is it considered acceptable to use mutable variables. For example, when using counters in a loop—but in many scenarios, even those can be handled by using immutable variables created with `val`.

Mutability and Type-Safety Caveats

There is another catch with mutable variables, and with this, we will now pivot toward type-safety in Scala. Let's take a look at it with the help of an example.

Create a mutable variable in Scala named `temperature` and then assign it a value of 98. As per the concepts developed so far, you can change its value, right? Try assigning it the value 100. It worked? Good.

Now try doing this—try assigning the value `"hot"` to the `temperature` variable. It's mutable and it should work, right? Refer to Figure 4-3 for further proof of this.

```
scala> var temperature = 98
temperature: Int = 98

scala> temperature = 100
temperature: Int = 100

scala> temperature = "hot"
<console>:12: error: type mismatch;
 found    : String("hot")
 required: Int
        temperature = "hot"
                      ^
```

Figure 4-3. *Mutability in variables created using the var keyword*

Surprise—it won't work.

It's not because of immutability. It's because Scala is a strongly typed language. This is yet another striking difference between Scala and dynamic languages like Python, which are pretty relaxed in this aspect. In Python, you can create a variable that may initially hold an integer value and in the next instant, it can hold a string or any other data type with no complaints.

In Scala, every variable that you create has a type strictly associated with it. For instance, if you create a mutable variable of type Integer, that type will remain associated with it. If you try to change its value to another integer, it will work. But if you try to change its value to another type, e.g., from Integer to String, it won't work. You saw proof of this in the previous example.

We cover data types in the upcoming chapter, but you have been using some of the types already—numbers (you assigned 98 to temperature) and strings (you assigned "Pakistan" to the country variable). So, in a nutshell, every Scala variable (mutable/immutable) must have a type and when it's initialized, the type gets associated with it and cannot be changed.

Specifying Types for Variables and Type Inference

There are two ways you can specify types of variables. This is where Scala is different from other statically typed languages like Java.

- Type inference

- Explicitly declaring type

By *type inference,* we mean that you let Scala infer the type of variable on its own. You don't dictate or specify the type, rather you let Scala make the best possible guess when you create a variable. That's exactly how we've been working with variables so far. Did we, at any point in time, specify that we wanted to create Int or String type variables? Nope. We simply issued expressions like this:

```
val temperature = 10
val country = "Australia"
```

In the first instance, Scala inferred the temperature type to be Integer and, in the second case, it guessed the type to be String. How can you be sure of that? One simple way is that in the Scala shell, when you create variables, right after the execution of that expression, it shows the type as well.

This characteristic is not linked to immutable or mutable variables. In both cases, it will infer the types whenever it can.

Alternatively, you can explicitly specify the type at the time of variable creation. In this case, you add another layer of emphasis/enforcement that this variable will always store the value of this type and nothing else.

The way to do this is to specify the type after the variable name separated by a colon :. Generally:

```
<val | var> <variable_name>:<variable_type> = <variable_value>
```

Here are some examples:

```
val country:String = "Australia"
var temperature:Int = 10
var isCustomer:Boolean = true
```

Generally, as a best practice, programmers explicitly specify the types to avoid any ambiguities or discrepancies. By doing so, you can be confident that you are working with the expected data types. Going forward, when you use functions, you will find that you specify the parameters along with their types.

This benefit will become even more evident when you use IDEs like IntelliJ because if the result of an expression is of a different type than what it expects, it will indicate an error. I know that we haven't covered IntelliJ yet, but I'm just giving you a heads up about this concept beforehand.

Scala Identifier Rules and Naming Conventions

So far you have learned how to create variables in Scala. Another important concept closely associated with variables is *identifier rules,* which generally refers to rules that must be followed while naming your variables (and other entities in Scala like classes, functions, etc.). Let's explore this further.

First, if you don't happen to follow the identifier rules, you will get errors while creating variables. To better understand this point, consider the following example:

```
scala> val 10years = 1
<console>:1: error: Invalid literal number
       val 10years = 1
         ^
```

At first, it may appear that the expression is syntactically correct. We used the `val` keyword, then added a variable name, and then tried to assign a value to it. However, we still got error. The error in this context is related to the variable name; Scala complained that it's an invalid literal number.

To avoid such errors, follow these basic identifier rules:

- All identifiers in Scala are case sensitive (meaning `firstname`, `FirstName`, and `firstName` are three different identifiers or variable names).

- An identifier must not start with a numeric digit (e.g., `10years` and `2Stores` are not valid identifiers or variable names).

- An identifier must not consist of Scala keywords (e.g., `def`, `class`, `for`, etc.). Also, An identifier name must not start with operators (e.g., `+`, `:`, `?`, `~`, or `#`).

In addition to these rules, it's always a good practice to follow naming conventions, which are a set of established naming standards. Variable names should be camel cased (in which the first word is lowercase and each subsequent word begins with a capital letter, with no intervening spaces or punctuation). For instance, `tableName`, `filePath`, and `regularizationParameter` are all valid variable names written in camel case. Whereas `TableName`, `Table-Name`, `TABLENAME`, and `file_path` don't follow that naming convention. Thus, whenever possible, make a habit of following these naming conventions in your code.

So in this chapter, you learned about the concept of the variable, its type (mutable/immutable), how type-safety works in Scala, and the rules governing their names. These concepts are fundamental to subsequent chapters, so make sure you understand them well.

EXERCISES

- Try creating a variable of type Double and assigning an integer value to it. Does it work? If yes, research this. Do it the other way (storing a Double value in an Integer variable).

- Try creating a variable (e.g., x) and assigning a value to it (e.g., 10). Then create another variable (e.g., y) and assign it to another variable (i.e., x=y). Now change the value of x. Check whether it changed the value of y or not. If not, research this concept (specifically what is meant by *pass by value* and *pass by reference*).

- Try creating multiple variables on one line.

CHAPTER 5

Data Types

In our daily life, we encounter data of varying nature. For instance, our names consist of letters, our mobile numbers are numbers, and our decisions are usually yes (true) or no (false). To represent these and many other types, every programming language has a type system to support this notion. Using the combination of different types that come with a language's type system, we can create variables associated with these types to implement tasks of varying nature. Like many other languages, Scala has a strong type system (which in fact is more sophisticated than many others) and there are different types available out of the box that you can use and work with.

Unlike other programming languages like Java, there is no concept of primitive data types in Scala. Primitive data types are considered the most basic data types like Integer, Boolean, Char, etc., and are just meant to store simple values of the type. In other programming languages, variables with primitive types actually store the value in them instead of storing reference to their addresses where they are stored.

In Scala, each type is actually an object. An *object* is an object oriented concept which, at a very high level, represents an entity (or in more simple terms a thing) that embodies fields and methods that you can call to carry out actions using that thing (we briefly covered this in Chapter 3). The notion of types in Scala is not as simple as it is in many other languages and that's the reason Scala has no primitive data types.

© Irfan Elahi 2019
I. Elahi, *Scala Programming for Big Data Analytics*,
https://doi.org/10.1007/978-1-4842-4810-2_5

Scala Type Hierarchy System

In addition to each type being an object, each type in Scala also belongs to a *type hierarchy,* as shown in Figure 5-1.

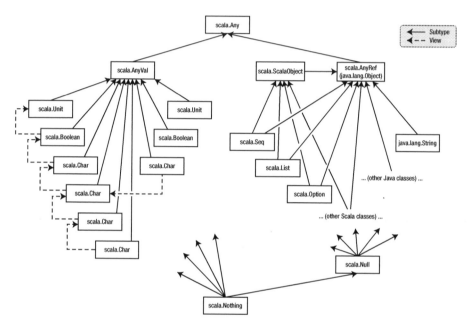

Figure 5-1. *Scala type hierarchy*

The hierarchy may appear complex at first, but let's look at it in step-by-step manner:

- At the top root of the hierarchy, there is a type called Any. This is the *parent* or *super-type* of all the types. All other types are descendants or children of the Any type.

- The Any type is further divided into two categories:

  - AnyVal: This is the parent of all value types. This is where the so-called primitive types fall:

- Numeric types: `Double`, `Float`, `Long`, `Int`, `Short`, and `Byte`. These types have hierarchies in them as well. For example, `Double` is the super-type of all numeric types and `Byte` is the lowest in this hierarchy of numeric types.

- Boolean type: Represent two values: true and false.

- `Unit`: Represents emptiness and is equivalent to void in other languages. We explore this in detail in this chapter.

- `Char`: Used to represent characters and is different from the `String` data type. In Scala, it's represented using single quotes `'`. Under the hood, they are stored as integer (unsigned) values.

- `AnyRef`: This is the parent of all reference types and is analogous to `java.lang.Object` in Java. Again, it's an object oriented construct and as of now, you may understand that when you define a variable of a reference type, the variable then stores the memory address of that variable instead of actually storing that value directly. When Scala is used in the context of the Java runtime environment, `AnyRef` corresponds to `java.lang.Object`. Similarly, when you create your own types in Scala, they will also fall under this category. The `String` type that you used briefly in the previous chapter also belongs to this category. This is also where all the Scala collections (such as `List`, `Map`, etc.) are found.

At the bottom of the hierarchy lies the Nothing type. In Scala, every expression that you write must return a value. Even those functions that display something (e.g., println) also in fact return data of type Unit. In such instances where your expression is not returning anything, Nothing is used. This is the case when there is a loop that runs infinitely or a function that stops an application.

Deeper understanding of this hierarchy system involves knowledge of object-oriented principles, as each sub-type (e.g., AnyVal is sub-type of Any) inherits certain methods from its parent. Also, the concept of polymorphism is closely associated with this type system. But at this stage, all you need to know is:

- The equivalent primitive types in Scala. That is, the numeric types in Scala, the non-numeric equivalent primitive types (boolean), etc.

- If you use a particular type in Scala, you should know at least where it falls in the overall hierarchy.

- If you find types like Any, Unit, etc., this shouldn't surprise you, as all of these are types, located at different levels in the overall hierarchy.

EXERCISES: DATA TYPES

- Understand the difference, also in the context of memory footprint and range, between different numeric types in Scala. What are the limitations of each and when does it make sense to use one over the other?

- Research what functions are available in the integer data type, i.e. addition, subtraction, multiplication, and division. Use them in Scala REPL.

- Research which operators on numeric types have precedence, i.e., in an expression, which operation will be executed before the other. Understand these basic concepts on your own.

In this chapter, instead of going into the details of all the types, I cover specific types in the Scala language, along with their respective implications, and will point you in the right direction to deepen your knowledge about Scala's type system.

The Boolean Type

In Scala, to show that something is true or false, we have the Boolean type. For example, to say is 2 less than 5 (which is true) and "Scala" is equal to "Java" (which is false), we use Boolean types. In this type, there can be only two values:

```
true
false
```

These values are case sensitive. True/False/TRUE/"True" are not equivalent Boolean values.

Boolean values generally appear when you are performing some logical comparisons, such as < (lesser than), > (greater than), >= (greater than or equal to), == (equal to). As an example, type the operator shown in Figure 5-2 into Scala REPL.

```
scala> val equalityResult = 2==1
equalityResult: Boolean = false
```

Figure 5-2. *Using a relational operator to generate Boolean values*

You'll get a variable of type Boolean whose value is false. The expression 2==1 returns a Boolean value (note that there are two = signs and not one).

Note   The == (double equals to) is a comparison operator that returns Boolean values, whereas = is an assignment that assigns a value on its right to the variable on its left. Also, as = is just an assignment operator, it does not return anything per se. However, in Scala, every expression normally returns something and thus in the case of simple assignment expressions, something does get returned and that's Unit.

EXERCISES: BOOLEAN TYPE

- Research the different types of logical operators available in Scala. Try to use them in Scala REPL.

- Try assigning a Boolean variable to an Integer variable. What do you get? Research whether you can do this in other languages.

- Try adding two Boolean values. What do you get?

The String Type

To represent alphanumeric data in Scala, string data is generally used. Specifically, if you enclose a value in double quotes (" ") or three double quotes (""" """), it becomes a string. As mentioned, a string can have numbers or alphabetical characters (or other characters if you are using encodings like UTF-8). In the context of type hierarchy in Scala, it's a subtype of AnyRef. The String type, even in other languages like Java, is not considered a primitive type, as it's a reference type. And in one respect, `String` is a collection of characters because you can access/ index individual elements of the `String` data. These concepts will become clearer soon.

You've already worked with a number of examples of `String` types but for practice, type the following:

```
val query = "Select * from table where id = 1"
```

In this example, the whole statement, starting from `Select` until `1`, is a string because it's enclosed in double quotes (`""`). It is stored in a `query` variable.

In certain instances, i.e. with a few special characters, you need to escape characters in your `String` variable. For example, if your `String` variable contains double quotes or a backslash, you have to proceed with \. If you don't, Scala will report back an error. For example, consider the example shown in Figure 5-3.

```
scala> val greeting = "my name is "Irfan Elahi" and I am from Pakistan"
<console>:11: error: value Irfan is not a member of String
       val greeting = "my name is "Irfan Elahi" and I am from Pakistan"
                                    ^
<console>:11: error: value Elahi is not a member of StringContext
       val greeting = "my name is "Irfan Elahi" and I am from Pakistan"
                                          ^
```

Figure 5-3. *The problem of not escaping certain characters in String*

If you escape these special characters with \, the issue is addressed properly, as shown in Figure 5-4.

```
scala> val greeting = "my name is \"Irfan Elahi\" and I am from Pakistan"
greeting: String = my name is "Irfan Elahi" and I am from Pakistan
```

Figure 5-4. *Escaping characters in String*

To elaborate the previous point that String is actually a collection as well and you can index/retrieve individual characters, try the following in your Scala REPL:

```
greeting(0)
```

See what is returned. You will find output similar to Figure 5-5.

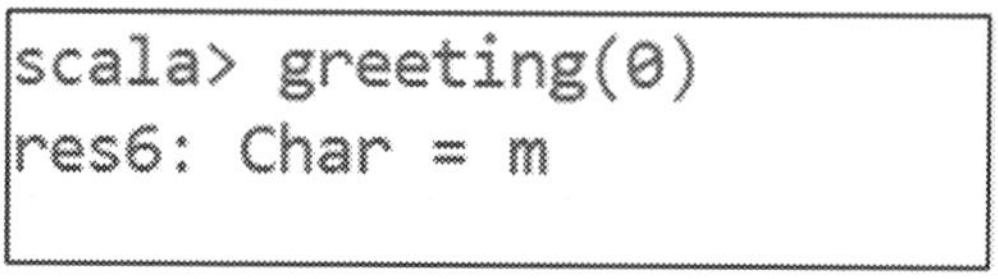

```
scala> greeting(0)
res6: Char = m
```

Figure 5-5. *Accessing individual characters of a String*

Although the concepts related to collection are covered in later chapters, you accessed the first element (or character) in the String data. The sequence of numbers in this context starts at zero. That is, if you access the 0th element, it will return the first character of your String, and so on.

Similarly, there are certain special characters as well:

- \n (newline character)

- \t (tab character)

- \b (backspace character)

- \r (carriage return character)

These special characters are used for specific purposes and I suggest you do further research about these to learn why and where they are used.

Multi-Line Strings

As mentioned previously, you can create a string in Scala with double or triple quotes. If your string consists of multiple lines or has quotes within itself, Scala gives you the option to use a triple-quoted string like so:

```
val funkyString = """ this is a
multi-line string and in this string,
there can be
"quotes" as well with
no problems"""
```

This results in the creation of a multi-line string.

EXERCISES: STRING TYPES

- Create a string variable and then type . (the dot character) and press Tab. You will see a list of functions. Many of them are covered in this book; however, explore them and learn what they do. The more you know about them, the better.

- Try converting numeric types and Boolean types to String types. Did you have any issue in doing so? You shouldn't.

String Operations

Let's delve into some of the common yet important string operations that developers use a lot in general purpose and Big Data applications.

String Concatenation

Try creating two string variables that contain numeric values:

```
val a="1"
```

and

```
val b="2"
```

and then do

```
a+b
```

What happened? Did it return 3 or 12? It should return 12. Why? This operation is called *string concatenation.*

String Interpolation

Try creating a variable as follows:

```
val name="irfan"
```

Then create another string variable as follows:

```
 val introduction = "my name is $name"
```

What do you get? You should get the following:

```
 "my name is $name"
```

This is straightforward, right? Now just append s as follows:

```
 val introduction = s"my name is $name"
```

Now what is the output? It should be this:

```
"my name is Irfan"
```

Why? That's how you do *string interpolation.* It will substitute the variable name with its value if a string starts with s. Similarly, give this a shot as well:

```
val introduction = s"my name is ${name}"
```

You will get similar output. You can do the same if you precede a string with f.

With string interpolation, there are some caveats that you should be aware of. To explore them, type this in Scala REPL:

```
val introduction = s"my name is $name.toUpperCase"
```

It will display the following:

```
"my name is Irfan.toUpperCase"
```

What you are trying to do is to use the `.toUpperCase` function, which comes with `String` to convert the `String` to uppercase, but you are unable to do that in the previous example. Now type this:

```
val introduction = "my name is ${name.toUpperCase}"
```

You should get this:

```
"my name is IRFAN"
```

Thus, if you want to use methods on your variables, like `.toUpperCase` in this example, you should use `${variable_name.method}`.

Also, as eluded to before, there are two ways you can do string interpolation in Scala—using `s` and `f`. The main difference between the two is that `f` provides an easier way to format the numbers. Consider the following example:

```
scala> val sharePrice = 100.4
sharePrice: Double = 100.4

scala> s"the share price is $sharePrice"
res0: String = the share price is 100.4

scala> s"the share price is $sharePrice%.2f"
res1: String = the share price is 100.4%.2f

scala> f"the share price is $sharePrice%.2f"
res2: String = the share price is 100.40
```

As you can see in this example, we first defined a variable of type Double that allows us to use decimal numbers. Then, using s for string interpolation, we tried to format the display of the number by enforcing it to use two digits in decimals (using %.2f). It didn't work with s, whereas it worked with f.

There is a raw interpolator as well, which you use by prepending the string with raw. When you use that, it uses the string as it is without any type of processing or interpolation:

```
scala> val aString="Irfan \n Elahi"
aString: String =
Irfan
 Elahi

scala> val aString=raw"Irfan \n Elahi"
aString: String = Irfan \n Elahi
```

In the first example, you create a String variable and use \n in the string value, which corresponds to a newline (remember that I referred to such types of special characters in this chapter before?). That's why in the output, you got Irfan and Elahi on two different lines. When you use the same String value and prepend it with raw, you see that it didn't process \n and just treated it in the raw form.

Length of String

As String can be thought of as a series/sequence/collection of characters, the notion of the length of String makes sense. In Scala, to get the length or size of a string, you can use the length or size method of the String types. For instance, to get the length of the following string variable, use this:

```
val customerPackage = "prepaid"
```

You can do this similar to what's shown in Figure 5-6.

```
scala> customerPackage.length
res8: Int = 7

scala> customerPackage.size
res9: Int = 7
```

Figure 5-6. *Finding string length or size in Scala*

Lastly, to expand more on string indexing, create a string variable:

```
val customerPackage = "prepaid"
```

Then type customerPackage(0). You will get p as was described previously (you'll get the first element). But if you try to access a number beyond the string's length, e.g. customerPackage(100), you will get an error indicating that you are trying to get an element out of its bounds/range (see Figure 5-7).

```
scala> customerPackage(100)
java.lang.StringIndexOutOfBoundsException: String index out of range: 100
  at java.lang.String.charAt(Unknown Source)
  at scala.collection.immutable.StringOps$.apply$extension(StringOps.scala:37)
  ... 28 elided
```

Figure 5-7. *Getting an element of string outside of its range*

Splitting Strings

Many times you need to split a string on a particular character to perform certain types of processing. For example, if you have a file that consists of comma-separated values (CSV) like this:

```
1,mark zuckerberg,facebook
```

It may represent a row from a database table and each value separated by comma may represent a column. In this example, "1" may represent some ID column, "mark zuckerberg" is the name of a person (another column), and "facebook" is a company (another column). If you have many records like these in a file and want to operate on one, a specific column/portion of this string (just the names), then you usually split the string into individual "components" or parts, as follows:

```
scala> val aRow = "1,mark zuckerberg,facebook"
aRow: String = 1,mark zuckerberg,facebook

scala> aRow.split(",")
res11: Array[String] = Array(1, mark zuckerberg, facebook)
```

As you can see in this example, when you use split, you define which character you want the string to be split on. In this case, we wanted it to split on a comma (,) so we specified that. Thus, wherever it encountered a comma, it performed the split and returned a different data type (Array[String], which is a Scala collection). Now when you use split, you get an array that holds all the different parts of the string. You can access the individual parts as follows:

```
scala> aRow.split(",")(0)
res12: String = 1

scala> aRow.split(",")(1)
res13: String = mark zuckerberg

scala> aRow.split(",")(2)
res14: String = facebook
```

Just a note that this scenario appears quite often when using Apache Spark, as CSV is a common data format, and you use Apache Spark to process large volumes of CSV files. The logic of split that you learned here is used in almost in the same way when using Apache Spark for large-scale processing.

Extracting Parts of a String

If you need to extract specific parts of a string between certain positions (or starting from a specific position), you can use the substring function of the String data type. Here's an example:

```
scala> val x = "apache spark"
x: String = apache spark
```

If you want to get the string starting from a specific position, you use substring as follows:

```
scala> x.substring(0)
res21: String = apache spark
scala> x.substring(1)
res22: String = pache spark
```

In this example, when you do substring(i), it returns all characters of the string starting at position I (whereas i starts at 0). Thus, if you use substring(0), it returns all the characters of the string starting from 0 (meaning all the characters because 0 is the starting point of a string). When you use substring(1), it returns all characters from 1 onward. That's why you got "pache spark" and it didn't return "a", which is the first character (at the 0th position).

You can use substring in another context to define a range of characters to be retrieved from a string:

```
scala> x.substring(1,4)
res26: String = pac
```

You passed two parameters to substring—substring(I,j). Specifically, you used substring (1,4) and it returned pac. It did so by returning characters starting at position i until j-1 (i.e., starting from position 1 and "p" exists at position 1) until 3 (i.e., 4-1 = 3 and the character "c" exists at position 3).

71

It's quite a handy tool if your data follows a specific format consisting of characters of a specific length (e.g., extracting month from a date, extracting ZIP codes, etc.).

Finding the Index of Characters in a String

If you want to find the position (also known as the *index*) of a specific character in a string, you can use the indexOf function of String, as follows:

```scala
scala> val x = "apache spark"
x: String = apache spark

scala> x.indexOf("a")
res27: Int = 0

scala> x.indexOf("p")
res28: Int = 1

scala> x.indexOf("k")
res29: Int = 11
```

You use the indexOf function by passing the character in its brackets whose position you want to find in a string. So in the previous example, when you pass use x.indexOf("a"), it returns 0, meaning the character "a" exists at the 0th position (which is the first position/index in Scala). Note that even though "a" appears multiple times in the variable, indexOf only returns its first occurrence.

This concludes the section on String data types. There are many functions available in String (and other data types) and it's not practically possible to cover all of them in this book. However, the ones covered so far will provide you with a firm foundation to develop your concepts/skills further.

Special Types in Scala

I will take this opportunity to discuss some special data types that are unique to Scala.

The Unit Type

Unit is a data type, just like Int, Double, String, etc. It generally appears when you use a function that doesn't return anything. I know we haven't used functions yet, but let's do one thing. In Scala REPL, type the following:

```
val printOutput = println("Hello Scala")
```

Assign it a variable (mutable or immutable) and then observe the output, which is shown in Figure 5-8.

```
scala> val printOutput = println("Hello Scala")
Hello Scala
printOutput: Unit = ()

scala>
```

Figure 5-8. *Data of type Unit returned by the println function*

Two main things happened here:

- You created a variable called printOutput and assigned it the result of println("Hello Scala").

- println is a function. You didn't create it, rather it comes by default with Scala. A function does something. println displays whatever you specify in () onscreen. It doesn't return anything. It just displays results. That's why, when you assign the result of println (which doesn't return anything), to a variable,

73

the type of printOutput is Unit. Why Unit? Again remember the rule in Scala that every expression must return something. Thus, when nothing is returned, to comply with this rule, a value of type Unit is returned.

You can think of Unit as empty. In other programming languages, the equivalent type is void. In contrast to this example, if you use a function that returns something, you won't get Unit:

```scala
val sqrtResult = math.sqrt(4)
```

The type of the variable sqrtResult will be Double because math.sqrt(4) returned a value (i.e., the square root of 4, which is 2).

Other scenarios in which you'll encounter Unit is when the last expression of your function is an assignment of a variable. More on that later.

Another way to think about the Unit type is when it's used, it means that you may be changing state of a variable (as a result of the assignment operation) which at times may refer to impurity, in the context of functional programming. This is not always the case though. Consider the following code:

```scala
scala> var aGlobalVariable = 10
aGlobalVariable: Int = 10

scala> def impureFunction() = {aGlobalVariable =
aGlobalVariable*2}
impureFunction: ()Unit

scala> aGlobalVariable
res8: Int = 10

scala> impureFunction

scala> aGlobalVariable
res10: Int = 20
```

In this example, you defined a variable (a mutable one using var). Then you defined a function (we haven't covered functions in Scala, so at this point, you can think of them as modules or pieces of code that you can write once and call multiple times). In the function, you are basically mutating or changing the value of the global variable. Note that the last expression of the function is the assignment expression and that's why the return type of the function is Unit. Then you displayed the original value of the aGlobalVariable and then called the function and displayed the value again.

Notice that the value has changed, which means that it's a side effect (you changed the value of a variable which was not in the body of the function). This is an example of an impure function and functions like these are generally avoided in Scala. This will be clearer when we cover functions in subsequent chapters.

When you use Apache Spark, there are certain functions that return Unit as well (e.g., foreach). In those instances, understanding Unit does help significantly.

The Any Type

In the Scala type hierarchy image shown in Figure 5-1, you may have noticed that the Any type exists at the root of the Scala type hierarchy.

How can you see it in action? Consider this specific instance. It appears when Scala encounters values of different types in a variable. For example, refer to Figure 5-9.

```
scala> var aList=List(1,"irfan")
aList: List[Any] = List(1, irfan)
```

Figure 5-9. *A list of type Any*

In Figure 5-9, you created a List. We haven't covered List yet but to elaborate on the concept, bear with me. Think of List as a type that can hold a collection of values. You can think of them as somewhat similar to String, as String holds a collection of characters. Normally, List in Scala contains elements of a specific type, for example, consider Figure 5-10, which is a list of integer values.

```
scala> val integerList = List(1,20,-100)
integerList: List[Int] = List(1, 20, -100)
```

Figure 5-10. *List of Integer type*

In Figure 5-10, Scala can infer the type right away—it's a List of Integers (as evident from the output of REPL i.e. List[Int]). But in Figure 5-9, it inferred a list of type List[Any]. Why? Because in Figure 5-9, the list was comprised of multiple types (Integer and String values). In this context, Scala fell back to a broader type (Any) to accommodate.

When you define a variable of type Any, it being the super-type of all types, it can house the value of any sub-type. For instance, consider Figure 5-11.

```
scala> var anyVariable:Any = 10
anyVariable: Any = 10

scala> anyVariable = "irfan"
anyVariable: Any = irfan
```

Figure 5-11. *Creating a variable of type Any in Scala*

Let's understand what's happening in Figure 5-11:

- You created a mutable variable of type Any.

- You stored an Integer value in it.

- You then stored a String value in it.

It didn't complain. Why? It's because it's a super-type or parent type of both Integer and String and had no issues in accommodating either of them.

That's where knowledge of type hierarchy comes in handy.

You may ask—is it a good practice to create variables of such "broad" types? The answer is that it depends, but generally the more specific a type, the better. For instance, if you are not sure about the value/type of a variable at runtime (i.e., when your program is actually running), in such instances it's safe to use the Any type as a catch-all type. Otherwise, it's recommended to use as specific a type as possible.

As you are working with data types, it will be helpful to note that in Scala, you can determine the type of an object by using the getClass method, as shown in Figure 5-12.

```
scala> sqrtResult.getClass
res15: Class[Double] = double
```

Figure 5-12. *Using the getClass method in Scala*

Type Casting in Scala

At times, you may need to change data from one type to another. Say you read user's input and you need to cast a number that the user entered as an Integer.

In Scala, to get user input, you use the following method:

`scala.io.StdIn.readLine`

When you use it, even if you type numbers, they are stored in the `String` type. If you want to perform numerical operations on the user's input, you need to cast the input as `Integer`.

In Scala, you do this using the `.to<Type>` functions. Figure 5-13 shows an example.

```
scala> val userInput = scala.io.StdIn.readLine
userInput: String = 800
```

Figure 5-13. *Getting input from a user*

When you use `scala.io.StdIn.readLine` and press Enter, it prompts you to enter input. Whatever you enter will be stored in the `userInput` variable. `scala.io.StdIn.readLine` is a function (like `math.sqrt` or `println`) and its return type is `String` (like the `math.sqrt` return type was `Double` and the `println` return type was `Unit`). So if you enter a number, it will be stored as a `String`. If for example you want to divide it by 10, you will get error, as shown in Figure 5-14.

```
scala> userInput/10
<console>:13: error: value / is not a member of String
       userInput/10
                ^
```

Figure 5-14. *Dividing a number in String form by another number*

You are trying to use the / operator (which stands for division for numerical types) on a variable that's a String type; thus it gives you an error. You need to convert your variable to a numeric type. To do that, you can use the toInt function, as shown in Figure 5-15.

```
scala> userInput.toInt/10
res17: Int = 80
```

Figure 5-15. *Type casting a String to an Integer type*

Similarly, you can do conversions to other data types.

This concludes this chapter, which covered the core aspects of data types in Scala!

EXERCISES: TYPE CASTING

- Try converting a Double (e.g., 10.5) to Int. What happens? It will drop the portion of number after the decimal. Beware of such nuances.

- Try running "10".toInt. Does it work? It should. Try to convert "two".toInt. Does it work? It shouldn't. You can't type cast all the time.

- Research how you generally work with nulls in Scala. You will find specific types, such as Option and its concrete subtypes (Some, None). Research them and make sure you understand their use.

CHAPTER 6

Conditional Statements

In life, we make a number of decisions at different moments. For example, if it's raining then we will not play outside; if a customer is susceptible to be churned, then we will use a particular marketing approach, and so on. There are many examples of conditional statements like this. Similarly, in programming, at different instances we have to consider a number of decisions and based on the results, decide the logic/flow of the program. For example, if a username exists and if the password matches, the user will be able to log in; otherwise, they are denied access. If we don't make these decisions, there can be no notion of intelligence in our programs and thus their usefulness is severely limited.

In programming languages, such scenarios are handled using *conditional statements.* In Scala, and in other languages, we use combination of `if/else` statements to handle tasks involving conditions. Using this construct, we can create a series of conditions to be checked and can define which actions to take (or which expressions to execute). These conditions can be nested if required to capture complex conditions. In Scala, there are certain nuances of using conditional statements and I highlight them as we go along. The goal of this chapter is to make you adept at using conditional statements in Scala.

© Irfan Elahi 2019
I. Elahi, *Scala Programming for Big Data Analytics,*
https://doi.org/10.1007/978-1-4842-4810-2_6

Furthermore, such constructs are heavily used in Big Data analytics. For instance, when you use Apache Spark to load data from external data sources (e.g., a distributed filesystem running on a set of machines), you can also commonly apply filter conditions to filter the records on which you will perform processing. For example, say you are loading a million lines of data from a series of text files and you want to process only those lines that contain a certain keyword (e.g. Scala). In that case, you will use similar constructs that will result in optimal processing.

Boolean Expressions

Before we dive into conditional statements, it will be helpful to revisit the notion of Boolean expressions.

A *Boolean* is a data type and has two possible values: `true` or `false`.

There are expressions in Scala that generate Boolean values, using certain operators. Just like certain expressions generate Integer results. For instance, if you type

```
1+10
```

You get `Integer` as a result. Similarly, if you type expressions that involve logical operators:

```
10 > 100
10 >= 100
"irfan" != "Irfan"
100 < 1000
100 <= 1000
```

All of these result in a Boolean value (`true`). All of these expressions use the following logical operators:

- < (less than)

- > (greater than)

- >= (greater than or equal to)

- == (equal to)

- != (not equal to)

Similarly, you can combine multiple logical expressions with operators, like so, in order to craft your desired conditions:

- & (AND)

- | (OR)

In the conditional statements that we'll cover next, we use these logical expressions extensively, so it's a good idea to practice them.

Using Conditional Statements in Scala

First, let's start with a simple example. We will define a variable called carBudget and apply a condition based on it. For example, if the car's budget is less than 30 (let's assume that the base unit is thousand dollars), we'll recommend buying a Mazda; otherwise, we'll recommend a BMW. This is a simple yet important decision that car lovers face in their pursuit of buying their dream cars!

First, enter the following in paste mode in Scala REPL (by typing :paste and pressing Enter):

```
:paste
val carBudget = 40
if (carBudget < 30)
println("buy Mazda")
else println("buy BMW")
```

When you execute it (by pressing Ctrl+D while in paste mode), you will get the following as output:

```
buy BMW
```

Step-by-Step Understanding of Conditional Statements

Let's look at how the conditional statement expressions that you wrote previously were executed:

- You define a variable called carBudget and assign it a value of 40.

- In the if statement and between the parentheses (), you specified a condition which was then checked, i.e., carBudget < 30, which turned out to be false (a Boolean value). As a result, the statements following the if statement didn't execute. Instead, the else block was executed.

Thus, from this example, you can gather the fundamentals of using conditional statements in Scala:

- You use if and else to constitute your conditions.

- The expressions in the if block are executed when the conditions defined in if () result in true.

- Otherwise, the else block is executed.

- The expressions that you write in the if parentheses must result in a Boolean value.

Caveats: Using {} After if/else

In the last example, we didn't use {} after the if or else statements. Our approach works correctly when there is only a single expression to execute after the if statement. However, there can be more than one statement in the if and else sections. Depending on certain conditions, you want to take an action that can be represented by a *series* of Scala expressions instead of just one. There are two ways to handle this in Scala:

- If the number of expressions in the if/else block is one, you can skip using {} after the if/else, like you did in the previous example. You can, however, still specify {} in such instances. So the previous example with {} can be written as follows:

```scala
val carBudget = 40
if (carBudget < 30) {
    println("buy Mazda")
}
else {
    println("buy BMW")
}
```

It would've worked fine as well.

- If the number of expressions in your if/else blocks are more than one, you must always use {} or you will run into logical errors. If you don't use {} and there are multiple statements intended to be executed after if/else, Scala will throw error. For example, try running this snippet:

```scala
val carBudget = 40
if (carBudget < 30)
println("so your budget is lesser than 30")
println("buy Mazda")
else print("buy BMW")
```

Scala will give you an error, as shown in Figure 6-1.

```
scala> :paste
// Entering paste mode (ctrl-D to finish)

val carBudget = 40
if (carBudget < 30)
println("so your budget is lesser than 30")
println("buy Mazda")
else print("buy BMW")

// Exiting paste mode, now interpreting.

<pastie>:5: error: ';' expected but 'else' found.
else print("buy BMW")
^
```

Figure 6-1. *Error when not using {} in if/else statements*

If you do this instead, it will work fine without any error (see Figure 6-2):

```
val carBudget = 40
if (carBudget < 30) {
    println("so your budget is lesser than 30")
    println("buy Mazda")
}
else {
    print("buy BMW")
}
```

```
scala> :paste
// Entering paste mode (ctrl-D to finish)

val carBudget = 40
if (carBudget < 30) {
println("so your budget is lesser than 30")
println("buy Mazda")
} else print("buy BMW")

// Exiting paste mode, now interpreting.

buy BMWcarBudget: Int = 40
```

Figure 6-2. *Correct use of {} in if/else statements*

So in a nutshell: get in the habit of using {} to surround the statements after the if and else blocks. That would be my advice.

Nested If/Else Statements

In order to represent conditions that are interdependent, you can nest if else conditions in Scala.

For instance, consider the following code:

```
val country="Australia"
val carBudget = 25
if (carBudget < 30) {
    println("So your budget is less than 30")
    if (country == "Australia") {
        println("Buy Mazda")
    } else {
        println("Buy Toyota")
    }
}
```

```
else {
    print("buy BMW")
}
```

Let's look at this code in a step-by-step manner:

- The first if condition checks if your budget is less than 30. If it is, the program flow goes inside the if block.

- Once in that block, there is another if condition that checks which country you belong to. If you live in Australia, it suggests you buy a Mazda; otherwise, it suggests a Toyota.

- If your budget is greater than 30, it skips the if block and executes the else block

So here you've used a nested if/else condition. A better way to visualize nested if/else conditions is via a decision tree like in Figure 6-3.

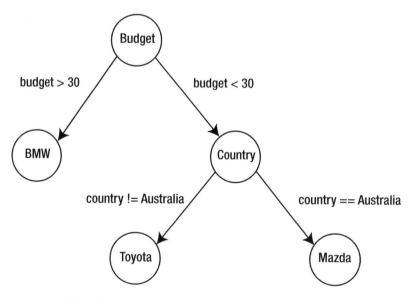

Figure 6-3. *Visual representation of nested if/else conditions*

We have not used conditions in the else block yet. It's an important construct so let's consider the following code snippet:

```
val carBudget = 70
if (carBudget < 30) {
    println("Buy Toyota")
}
else if (carBudget > 30 & carBudget < 50) {
    print("Buy Mazda")
} else if (carBudget > 50) {
    print("Buy BMW")
}
```

In this code example, notice that the code first checks whether carBudget is less than 30. If it's not, it checks another condition in the else block. If it's greater than 30 and less than 50 and if that condition is not met, it checks the last else block's condition. This way, you can write more involved conditional statements to achieve your desired flow of conditions.

If/Else as a Ternary Operator

A *ternary* operator is a way to express conditions in a concise way and is generally available in different languages (even in Excel). As you've seen previously, if/else blocks span multiple lines of code. If your conditions and the corresponding actions are simple enough to be expressed in one line, for the sake of brevity, you can use ternary operator notation to create your conditional statements.

The form generally looks like this:

```
if <condition> <value to be returned upon True condition> else
<value to be returned upon False condition>
```

You can use a ternary operator in Scala using if/else as follows:

```
scala> val salary = 95000
salary: Int = 95000

scala> val highlyPaid = if (salary>100000) true else false
highlyPaid: Boolean = false
```

The example uses an if/else statement in one line without using blocks or multiple lines. If you have conditions that can be expressed conveniently in one line, you can use if/else in this way.

Pattern Matching

One of the reasons that so many developers love Scala is because of its powerful pattern-matching feature. Pattern matching is fancy term that basically means using match and case statements and the features that this combination supports.

If you have used other programming languages like C or Java, you may recall that they have a switch operator that allows you to check multiple conditions. You can do the same with multiple if/else statements, but switch provides better and cleaner syntax. You can achieve the same effect, with many capabilities, using pattern-matching constructs, as follows.

```
scala> :paste
// Entering paste mode (ctrl-D to finish)

val country = "Australia"
country match {
case "Australia" => "Continent"
case _ => "Not Continent"
}
```

```
// Exiting paste mode, now interpreting.

country: String = Australia
res9: String = Continent
```

Here's a step-by-step explanation of the code snippet:

- You first defined the variable called country and assigned it a value called "Australia".

- You then used the match keyword to match the value of the variable and created a code block within the {}.

- Within the code block, you wrote a series of case statements, which will act like conditions. The syntax of the case statements in the most basic form is:

```
case <condition> => <value to be returned>
```

By using case "Australia", you are basically doing an equality check; you are comparing whether the value of country is equal to "Australia". If the condition is true, the code to the right of => will be executed. In this case, the condition is true and there is only one expression to the right of =>, "Continent", so it will return that.

You used another case statement to check for another condition. It's like using another if/else condition. In the second case statement, you used _. You didn't specify a value to match. When you use _ in this context, it means that everything else will be matched. In other languages, the equivalent construct is default when you use the switch statement. So if nothing matches, case _ will act as a catch-all and will get executed. Similarly, when this case statement gets executed, whatever is to the right of => will be executed and ultimately be returned. In this case, just the false will be returned.

So you achieved a conditional flow using pattern matching.

Pattern matching is a huge topic and it's heavily used by developers to structure the conditional flow of their programs. Let me highlight another way you can use pattern matching in Scala:

```scala
scala> :paste
// Entering paste mode (ctrl-D to finish)

val salary = 95000
salary match {
case x if x>100000 => true
case y if y<100000 => false
}

// Exiting paste mode, now interpreting.

salary: Int = 95000
res8: Boolean = false
```

In this example, the overall concept remains the same—you are matching the value of the salary variable and specifying a couple of case statements, which act like conditions/checks. Depending on which case statement's condition becomes true, its code to the right of => will be executed.

But here's a subtle difference: you may have noticed that we used case statements a bit differently here:

```scala
case x if x<100000=> true
```

In this case, we used a variable x and then if statement within it. The value of the variable to be matched, i.e. salary, is passed to x and then the condition is checked (i.e., if x>100000). Depending on the result of the condition, the corresponding blocks are executed. If you want to perform such types of logical or conditional checks using pattern matching, that's how you do it—you use a variable, in this case x, and then apply logical conditions to it. Previously, when you didn't use this construct, you were merely checking the equality of variables. But to perform more involved conditions, you can use this form.

The scope of the variable in this case is limited to the case statement and the code body to the right of =>. You can use the x variable in the code block that you can write to the right of =>. But you can't use that variable x anywhere else, even in another case statement. Beware of this caveat. Using this variable in this way allows you not only to check conditions but also to use that variable in the code block on the right, which can be quite handy at times.

Similarly, in the second case statement, you defined another variable called y and applied a condition on it.

This concludes this chapter on conditional statements in which you mainly explored the concepts related to Boolean expressions and operators, using if/else blocks in your code, and using pattern matching to introduce a certain degree of decision-making in your programs. These constructs are heavily employed not only in general-purpose programming but also in Big Data analytics context. So be sure to practice the concepts as much as you can.

EXERCISE

- First, create a decision tree indicating a different series of conditions to be checked. Then use the if/else condition statements based on this decision tree in your program.

- Try assigning the whole if/else blocks to a variable and seeing what value is returned.

- Try nesting different case statements in pattern matching. See how it works.

- Explore other use cases of pattern matching. Explore how it can be used with regular expressions, type checks, and for catching exceptions.

CHAPTER 7

Code Blocks

As you progress through this book, you will come to appreciate that Scala embodies a number of constructs that you can leverage to your advantage. Such constructs help structure your code better, reduce verbosity, and improve productivity. In this chapter, we explore a feature in Scala that will help you achieve some of the aforementioned benefits.

The feature we will cover in this chapter is called the *code block,* which basically allows you to write a bunch of statements together in a block and assign the result to a variable. The block is processed as a unit and the expressions within that unit are executed within that scope. The result of the last expression of that block will be returned and will be stored in a variable.

This has a number of applications, but mostly it's used to assign values to variables after doing some preprocessing.

Code Blocks in Scala

Let's look at code blocks using a concrete example. Consider the following code:

```
val resultOfBlock = {
    val a=2
    val b=2
    a+b
}
```

Let's examine this code in a step-by-step format:

1. First and foremost, to create a block, you use {}. All
    the expressions between {} constitute a block and
    will be evaluated as a block.

    •   You created a variable called `resultOfBlock` and
        then assigned it to a block (a set of expressions
        within {}). The block included three statements.

    •   In the first statement, you created a variable
        called `a` and then created another variable `b`. Then
        you summed them using `a+b`. The result of this
        expression is 4. Also, this is the last statement in
        this block, so it will be returned and stored in the
        variable `resultOfBlock`. That's why the value that
        gets stored in the variable is 4.

This concept is further highlighted in Figure 7-1.

```
val variableName = {
// multiple expressions can be
//written here
```

```
}
```
The last expression will determine what gets
returned and stored in the variable

Figure 7-1. *Highlighting the concept of code blocks in Scala*

Note that you didn't explicitly type any `return` statement. You wrote
the block in such a way that the last expression resulted in 4 and thus it was
stored in the variable.

Another way to understand code blocks is that they allow you to encapsulate processing that needs to happen before you assign the final value to a variable. You might argue that you could also use functions for this purpose, because they allow you to do the same. And you could. However, functions make more sense if they are reusable; i.e., if you want to execute the same set of expressions again and again, you can create a function out of them that you can call repeatedly. If there are instances where you find that the processing/expressions won't repeat, then you can use code blocks.

Are code blocks necessary? Nope. You can achieve these tasks without them. It's just that they impart a structure to your program and improve readability.

I've used them extensively. For instance, in one application, I needed to use a file path passed to the program at runtime, validate that the file path existed, load the contents of that file, and then parse that file as JSON, and ultimately store the contents in a variable. I conveniently used a code block and it did the job well. Don't worry if some of the terms in this example don't make sense, as the intent was to convey a real-life example of code blocks use.

Caveats of Code Blocks

Like many other constructs of Scala, you must be cautious of certain caveats associated with code blocks. Let's look at those caveats through a number of examples.

In your Scala REPL, execute this block:

```
val resultOfBlock = {
    val a=2
    val b=2
    val c=a+b
}
```

What is the value of `resultOfBlock`? It will be `Unit`. Why? To answer this, look at the last expression of the {} block. It's as follows:

```
val c=a+b
```

As mentioned in the previous chapters, when you use the assignment expression, the return type of that expression is `Unit`. Assignment of a value to a variable does that task, but from a "value returning" point of view, it returns `Unit`, aka void, aka emptiness. So beware of that.

The same effect will be achieved if your last expression is a `println` in your block, or any function that doesn't return anything (or returns `Unit`). Go on and give it a try.

Thus, you can infer that the value returned from a block is governed by the last expression in that block.

Code Blocks and If/Else Statements

Now I may have mentioned it before, but I am stating it again—every statement that you type in Scala is an *expression*, which means it will return something. It doesn't matter if it's one line (1+2) or multiple lines, as is done in a code block.

Following this concept, `if/else` statements, which you learned about in Chapter 6, are also expressions. Thus, following the same principle, they will also return a value.

Now this is where things get interesting, because there are certain nuances that deserve further consideration.

Consider the following `if` statement:

```
val age = 50
val isOld = {
    if (age>50)
    true
    else false
    }
```

Consider if you execute this in Scala REPL, as shown in Figure 7-2.

```
scala> :paste
// Entering paste mode (ctrl-D to finish)

val age = 50
val isOld = {
    if (age>50)
    true
    else false
    }

// Exiting paste mode, now interpreting.

age: Int = 50
isOld: Boolean = false
```

Figure 7-2. *Highlighting the use of if/else statements and code blocks together*

Let's look at a number of things that happened in this code example:

- You are using if/else conditions like before. You are already familiar with them, so this is nothing new here.

- You have surrounded the if/else statement in a code block. How? By using {} around them. This means that last expression of this block will be returned and you are storing that to-be-returned value in a variable called isOld.

- Now, depending on the value of the age variable, either the if or else block will execute. In our example, because the age variable's value is 50 and it's not greater than 50, the else section is executed. As a result of that, false is returned.

- Try setting age to a lower value, e.g. 30, so that the if block is executed. Check what value is stored in the isOld variable. It should be true.

- Furthermore, note that in this example, the type of the isOld variable is Boolean. This makes sense because no matter which if or else block gets executed, it will always return a Boolean value (true or false), right? Now what if things aren't that pure and you do something like this:

```scala
val age=50
val isOld = {
    if (age>50)
    100
    else "no"
}
```

Upon execution, you'll get something resembling Figure 7-3.

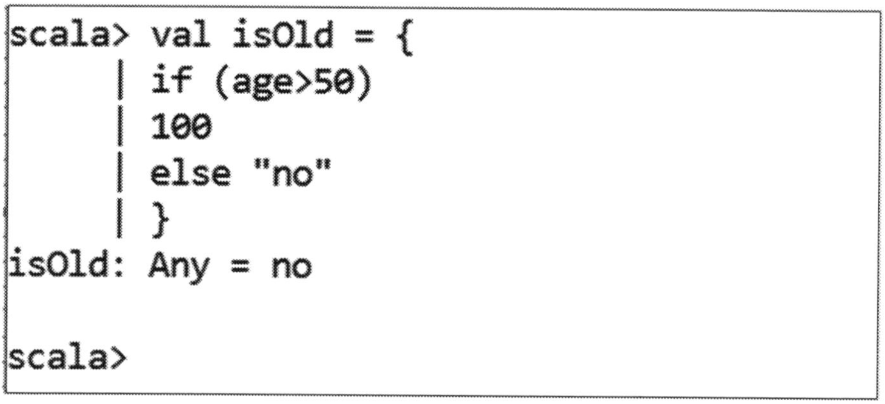

```
scala> val isOld = {
     |  if (age>50)
     |  100
     |  else "no"
     |  }
isOld: Any = no

scala>
```

Figure 7-3. *Highlighting the use of if/else statements and code blocks together*

You will find that the type of the isOld variable now is Any. It's not a string like before. Why?

Let's focus on the if block. What value is there? It's 100, which is an Integer. And what value will be returned in else? It's "no", which is a String. So your if/else consists of two possibilities for returning values: an Integer and String. It can return two different values belonging to different types. This scenario is different from the previous one, where the contents of the if and else blocks were of the same type. In the current scenario, based on Scala's type inference characteristic, it inferred the type of this variable to be Any, which is a super-type (remember the Scala type hierarchy?).

Beware of such nuances when you use code blocks. They are effective and powerful but come with such caveats if not handled properly.

EXERCISES

- Practice using code blocks by writing a series of statements in them and assigning a value to a variable to see what is returned and stored.

- Study the caveats of what value gets returned when you use pattern matching, i.e., if you return different values in different case blocks, make sure you understand what happens.

CHAPTER 8

Functions

When you are programming, you are solving a specific problem. The nature and complexity of that problem can vary. It can be as simple as finding a square root of a function or it can be as complex as writing data cached in RAM to multiple nodes in your environment in a parallel fashion (hint: Apache Spark). In either case, you write code and expressions that help you address that problem in your program. When you are in the mindset of programming, you think about what your input is and what your output is. In the case of finding the square root of a number, the number is your input and the square root of that number is your output. Then you write statements that help you address that problem. This concept lays the foundation of functions that we will explore in detail in this chapter.

Why Use Functions?

What if you need the same functionality multiple times in your program? Let's say it's needed at the start of a program, somewhere in between, in some other file that's part of your overall project, and at the end. You know how to find the square root. But how can you use this approach multiple times in more than one place in your application?

Like many things in life, you have a choice to make here:

- You could rewrite all the expressions again and again whenever you need to find the square root of a number.

© Irfan Elahi 2019
I. Elahi, *Scala Programming for Big Data Analytics*,
https://doi.org/10.1007/978-1-4842-4810-2_8

- You could write it once somewhere in your application in a way that's easily accessible and useable quickly whenever you need it. In other words, you can reuse the expression whenever and wherever needed.

Let's take a pause and think about which approach is the better one and why it's better.

With the first approach, you are rewriting the same piece of code again and again, and you are writing it in multiple places. This will increase the cost of maintaining your code. If you need to fix or improve your code or want to introduce customization to finding the square root (e.g., handling exceptions when users try to find the square root of a string), you would have to sift through dozens of places in the code to make the change (provided you are able find them all in the first place!). Also, while making changes, there is always a possibility that bugs will be introduced of varying nature (e.g., syntax errors or logical errors). Additionally, if the function behavior is intended to remain the same no matter where and how many times it's used, how will you test that the different pieces of code scattered all around your application are behaving/working identically?

In the world of programming, the practice of duplicating code is considered a sin! There is a specific phrase that's quite common with the evangelists of programming—DRY, which stands for "Don't Repeat Yourself". Whenever possible, avoid rewriting your code. Whenever possible, maximize the reusability of your code. Whenever possible, abstract the internals of your code from the external world or the program that uses that functionality.

In view of all the recommendations stated here and all the bad things that happen if you ignore them, you can deduce that good programmers use functions!

Functions are pieces of code (e.g., finding the square root of a number) wrapped in such a way that you can use them whenever you need them, without having to write the actual logic again and again. You define the logic of a function once and then just call that function whenever you need it.

Additionally, you rely on parameters that are passed to functions that render functions dynamic. Parameters are variables that are used in functions and, when calling functions, you can specify their values. The function uses those values in the core logic to generate output. In our square root example, the parameter would be the number from which we want to find the square root.

Also, when you test your code, functions become even more important as you structure your code in the form of functions that you can individually test. If your code's logic is not comprised of functions, it becomes difficult to test your code, and this can lead to a number of challenges when you want to make and test changes in your code.

Thus, in summary, functions are recommended in programming. Particularly in Apache Spark programming, functions are used extensively and this notion will become evident when we cover Scala collections.

Understanding Functions

As a best practice, it's recommended that your functions be focused and smaller and should be designed in such a way that they do one and only one task. If your functions are doing lots of tasks and are becoming complex and lengthy, it's probably a good idea to break them into smaller pieces. If you know what a function is doing, it will always help when you test its functionality.

Furthermore, before we delve into the hands-on part of this chapter, it's recommended that functions not generate any side-effects. Functions should operate only within their scope. They should operate only on the parameters passed to them. They should not tamper with or alter the state of any variable defined outside of their scope. These recommendations are biased in the light of functional programming, but the more you follow this rule, the more robustness there will be in your code. You won't have to track the impact of your functions outside their scope. It makes life easier.

Functions in Scala

With that understanding, rationale, and motivation behind using functions established, let's dive into the actual hands-on stuff—how to develop functions in Scala!

In Scala, the syntax for defining functions looks like this:

```
def <function_name>(<parameter_name:parameter_type>,
<parameter_name:parameter_type>…):return_type = {
  //function body
}
```

Let's decipher this syntax:

- Function definitions in Scala start with the def keyword.

- Every function has a name (function_name in the previous example).

- A function can accept zero or more parameters. Those parameters are defined within the parentheses () after the function name. The parameters are strongly typed, which means you have to explicitly specify their types. Think of parameters as variables of functions and these variables are available for your use in the function body.

- These parameters are not available outside of the function.

- Every function has a return type, which can be denoted by :return_type after the params. This means that every function returns a value of some sort whenever it is called. You can skip specifying the return type. If you do, the return type of the function will be determined based on the last expression in the function body.

- The body of the function is encapsulated within brackets {}. If it's a single-line function, you don't have to use the brackets. More on this later.

Here is a concrete example:

```
def getSquareRoot(givenNumber:Int):Double = {
    println(s"Finding square root of $givenNumber")
    math.sqrt(givenNumber)
}
```

Let's look at this function in the step-by-step way:

- You defined a function with the name getSquareRoot. The name can be any valid identifier, as we discussed in Chapter 4. Also, as per Scala's naming convention, the name should start with a lowercase letter. Generally, function names represent verbs, as they are meant to do something.

- This function accepts or expects one parameter defined within the (). Also, it further expects the function to be a specific type. In this case, it's going to be integer, as you are going to find the square root of an integer number.

- This function, after it does what it's supposed to do, which is find the square root in this case, returns a number of type Double.

- You then specify what happens inside the function or what goes in the body of the function. That's where you specify all the logic of that function and implement the functionality for which the function exists.

107

Figure 8-1 further elaborates these concepts.

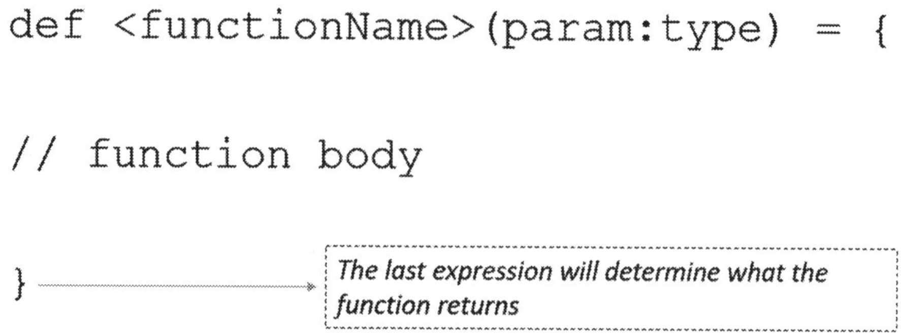

```
def <functionName>(param:type) = {

// function body

}
```

> The last expression will determine what the
> function returns

Figure 8-1. *Illustrating the concept of functions in Scala*

That's it. That's how you usually define functions in Scala. It's important to learn this syntax convention in Scala, as you are going to extensively define functions in Scala.

Invoking a Function

Up until now, you have defined a function and implemented the logic of finding the square root of the number that you pass to it as its parameter. Defining a function is one half of the puzzle. Defined functions are useless as they won't execute on their own unless you actually use/invoke them.

Let's look at how to invoke this function.

In Scala, invoking a function is pretty straightforward:

```
scala> getSquareRoot(25)
finding square root of 25
res9: Double = 5.0
```

You use the name of the function and then pass any required parameters. If you don't pass parameters where the function is expecting them, the compiler will throw error, as the function will have no value to work with.

If a function is returning a value, you can assign the function to a variable as well:

```scala
scala> val squareOf25 = squareThis(25)
```

Caveats: Function Definition

Let's consider a couple of caveats worth mentioning when it comes to defining functions in Scala.

Type Inference

You can skip specifying the return type of a function if you want and leave it to the compiler to figure out what the return type of your function will be. If you follow this approach, you can define the same function that you defined before in an alternative way, as follows:

```scala
scala> def getSquareRoot(givenParam:Int)= {
    println(s"finding square root of $givenParam")
    math.sqrt(givenParam)
    }
squareThis: (givenParam: Int)Double
```

It will still work. In this example, we didn't write the return type (Double) when defining the function, as seen here:

```scala
def getSquareRoot(givenParam:Int):Double
```

Also, based on the output, you can see that Scala figured out that your function is returning a Double value.

Generally, it's recommended that you specify the return type of your function, as it enforces type-safety and improves readability of your code, which is equally important.

Return Statements: To Use or Not To Use?

If you have worked in other programming languages like Python, you may have an idea that in such languages, you explicitly use a `return` statement in your function to signify what will be returned from your function.

Did you use a `return` statement in your function body? You didn't. Do you remember what code blocks are and how they work? If not, go back to Chapter 7 and read about them. (Just a hint: In code blocks, the returned value was the last expression of the block.) In the case of a function, the same holds true—the last expression of a function is returned from the function or determines the return type of that function.

Referring to the example that we've been using so far: What was the last statement in the `squareThis` function? It was `math.sqrt(givenParam)`, thus the result of that expression, which was a number of type `Double`, will be returned.

Try changing the order of the expressions in your `squareThis`—make `println` the last expression of your `squareThis` and try to run the program. What happens? Your return type will become `Unit`. Why? Because the last expression is now `println`, whose return type is `Unit`. If you are wondering about the return value of your function, look at the last expression of your function body. There lies the clue!

Also, try using `println` as the last statement of your function with the return type highlighted as `Double` in the function definition. See if the Scala REPL allows you to do that. (Hint: It won't. Why? Because you enforced type-safety by specifying a return type in the function definition, whereas if you don't, your function will not care what the return type is, as it will solely depend on the last expression of your function's body.)

Functions with Multiple Parameters

You know that your function has one parameter and you just invoked it:

```
squareThis(5)
```

The value 5 is assigned to the `givenParam` variable that you defined in the function.

What if your function has (or requires) more than one parameter? Say you want to greet a company employee like so:

```
Hi Irfan. Welcome to Facebook!
```

How can you do this? You'll use two parameters like this:

```
scala> def greetEmployee(name:String, company:String) = {
    println(s"Hi $name. Welcome to $company")
}
greetEmployee: (name: String, company: String)Unit

scala> greetEmployee("Irfan","Facebook")
Hi Irfan. Welcome to Facebook
```

Let's dissect this further:

- Like before, you defined a function with a valid name, in this case `greetEmployee`.

- In the parentheses ( ), you specified multiple parameters. In fact, two of them—`name` and `company`.

- While invoking the function, you passed not one but two values—`"Irfan"`, `"Facebook"`. Why two? Because you had two parameters defined in your function definition. Try passing one parameter and see what the Scala REPL says.

111

- What is the return type of this function? It's Unit. The last statement in your function body is println and your last expression/statement determines what is returned. What does a println function return? Unit.

Positional Parameters

You might have noticed that the first argument value that you passed—"Irfan"—was assigned to the name parameter in the function definition.

Note   When you define functions, the variables in the function definition are called *parameters* (e.g., in the greetEmployee function, name and company are parameters). However, when you invoke that function and pass values to those parameters, those values are called arguments. You passed the "Irfan" and "Facebook" arguments.

Similarly, "Facebook" was assigned to the company variable and you used them in your function body to do something (in this case, to print a message). So it's like a positional assignment. The first value was assigned to the first variable, the second value was assigned to the second parameter, and so on.

Scala provides another way to pass arguments to functions, wherein you can specify the name of the variable as per the function definition and then assign values to it.

For example, you can use the greetEmployee function in the same way but invoke it in a slightly different way as follows:

```scala
scala> greetEmployee(name="Irfan",company="Deloitte")
Hi Irfan. Welcome to Deloitte
```

Have you noticed in this example how we passed argument values while invoking the function? We used the `parameter variable name = parameter value` format. When you do that in this way, you can change the order in which you pass arguments. You can specify the `company` parameter first and `name` second and Scala will understand it.

```scala
scala> greetEmployee(company="Deloitte",name="Irfan")
Hi Irfan. Welcome to Deloitte
```

If your function has a lot of parameters (which generally isn't considered a good design decision), it's handy to invoke a function the way I specified, where you explicitly specify the parameter and parameter value instead of relying on the positions. One of the reasons to have fewer parameters is that it improves the readability of your code. Readers will clearly know what values are being passed to which parameters. Cleaner code also mitigates potential errors, as you'll have a better idea as to which argument values to pass to which parameters, and in what order. Code readability is a metric that you should always keep in mind while writing your code.

Default Value of Parameters in Functions

In certain instances, it helps to specify a default value of function parameters. If your function has a number of parameters and there are certain parameters whose values are unlikely to change and you can work with their default values, it's possible in Scala. Another way to interpret the benefit of this approach is that if you don't want to pass values to function parameters all the time and you think the default values of such parameters will work in most cases, it helps to use this feature because if you don't provide values for those parameters, Scala won't complain. Rather it will happily use the default values.

Going further into detail, in such a scenario, there are two possibilities:

- If you don't pass a value to the parameter that has a default value defined, Scala picks up the default value.

- If you pass a value to such parameters while invoking functions, Scala will use the value that you pass during invocation. In other words, the passed value will override the default values.

Here's an example to illustrate this concept further:

```scala
scala> def greetEmployee(name:String="Irfan", company:String) = {
    println(s"Hi $name. Welcome to $company")
}
greetEmployee: (name: String, company: String)Unit
```

In this code example, you've defined a function with the name parameter and it has a default value of Irfan. Cool. You can invoke this function as follows, whereby you don't specify a value for the name parameter:

```scala
scala> greetEmployee(company="Deloitte")
Hi Irfan. Welcome to Deloitte
```

Did you notice that it picked up the default value of the name parameter? To override the default value of the name parameter:

```scala
scala> greetEmployee(name="John Doe",company="Deloitte")
Hi John Doe. Welcome to Deloitte
```

It picked the value that you specified while calling the function (i.e., "John Doe" instead of the one that you used as the default in the function definition (i.e. "Irfan").

Function with No Arguments

There can be scenarios in which your functions don't require an argument (known as *zero parity*). In such scenarios, you can skip specifying the parentheses after the function names, like this:

```scala
scala> def printDate = java.time.LocalDate.now.toString
printDate: String

scala> printDate
res5: String = 2018-08-19
```

Here, we used the `java.time` library to print today's date. Don't worry if you haven't used this library before, as the focal point here is that the function didn't require any parameters. You can see that in the function definition (i.e., `def printDate`), I didn't use parentheses.

Another point to note is that when I invoked this function, I also didn't use the parentheses. In our previous examples, whenever we invoked functions, we invoked them with the parentheses. Here, as the function doesn't have any parameters, you can type the `printDate` without parentheses to invoke it.

Single-Line Functions

You must have noticed that we surrounded the body of functions within brackets {}. However, if your function's body occupies a single line instead of multiple lines, you can omit the brackets that usually surround the function body, as I've done in the previous example. But do note that you must use the brackets if your function body occupies multiple lines.

As a matter of personal preference, I always use the brackets, whether a function occupies a single line or multiple lines. That way, if there is any need to modify the function in the future and that modification results in the function occupying more than one line, I don't have to worry.

115

Using the return Statement in Functions

So far, I've highlighted that when using functions in the Scala language, you don't need a `return` statement to signify what value will be returned from a function. This is in contrast to many other languages like Python, which expect you to explicitly write `return` statements. There is one particular scenario where Scala does expect you to write `return` statements or you'll get an error.

If you've studied programming before, you may be aware of the idea of *recursion,* whereby a function repeatedly calls itself to perform a specific task and each subsequent call makes the function converge to a base-case (I advise to research recursion, as this isn't in the scope of this book). When you use recursion semantics, it becomes imperative to use `return` statements in such functions.

After covering many caveats of using functions in Scala, let's buckle up and explore some advanced concepts that are entrenched in Scala's functional programming roots.

Passing Functions as Arguments

When I say that Scala is a *functional* programming language, this shouldn't sound like Greek to you at this point. You must be able to recall functional programming language concepts. Functions are first class citizens; they can be treated like any other object.

So far, you've passed parameters of specific data types to functions. Those data types can belong to Scala's type hierarchy or they can be your own classes. Also, prior to the dawn of functional programming, the very idea of passing a function as an argument to other function was nonexistent. Given the fact that you are using a functional programming language, you can pass functions as parameters to functions. Trust me, this is quite a powerful concept.

You will see this concept leveraged a lot when you use functions like map, foreach, etc. on Scala collections. Many of these functions are extensively used in Apache Spark APIs in almost the same syntactical form. So learning these concepts now will have long-term benefits when you do Apache Spark development.

To drive the discussion forward, here's an intuitive example to explain this concept. You want to formulate a single generic function that will be responsible for handling both of the following cases, depending on the type of arguments (functions in this case) passed to it:

- It can convert strings to lowercase

- It can also convert strings to uppercase

How can you approach this in Scala, leveraging the functional programming constructs?

Let's create implementations of these two cases as follows, by creating two functions:

```scala
scala> def convertToUpper(name:String):String = name.
toUpperCase
convertToUpper: (name: String)String
```

```scala
scala> def convertToLower(name:String):String = name.
toLowerCase
convertToLower: (name: String)String
```

You've created two functions that do these jobs—one converts a string to lowercase and the other converts it to uppercase.

Let's now create a higher order function that will be responsible for handling both of these cases or responsible for calling these two specific functions.

You define a function like this:

```
scala> def changeCase(givenName:String,caseConverter:(String)=>
String) = caseConverter(givenName)
changeCase: (givenName: String, caseConverter: String =>
String)String
```

Let's spend some time dissecting this code example:

- You defined a function with the name changeCase. Its
  first parameter (givenName) is of type String and this
  is the string that you will convert to either upper or
  lowercase.

- Things get interesting if you observe the second
  parameter. Firstly, the parameter name is caseConverter.
  Then, after the colon :, you usually specify types. In this
  code example, you specified the "type" this way:

  (String)=>String

What does this mean? In Scala, it means any function that can accept
String as a parameter (signified by (**String**)=>String, i.e., the one before
=>) and can also return String (signified by (String)=>**String**, i.e., the
one after =>).

This is a generic representation, as there can be many functions that fit
this role (i.e., that can accept one String parameter and return a value of
type String).

You can also think of it as a wrapper that can hold any function that
fulfills two conditions: accepts one parameter as a String and returns the
String type.

Now compare that to the two functions that you created before—
convertToLower and convertToUpper. What do they accept? One String
parameter. What do they return? The String type. So they can be used here
as well. They can be passed to that second parameter (caseConverter)
that accepts functions of the form (String)=>(String).

Consider the body of the function as well. You invoke the `caseConverter` function and pass the parameter (`givenName`). The invoked function does its job (depending on the function you passed to the `caseConverter` parameter) and returns the results, which are then returned by the outer function (`changeCase`).

So to invoke such a function, you do something like this:

```
scala> changeCase("irfan",convertToUpper)
res12: String = IRFAN
```

In the first parameter value, you passed the `"Irfan"` string. That's a no brainer by now. For the second argument, you passed the name of the function that you created before—`convertToUpper`. This function will be passed to the `caseConverter` parameter. Now, `caseConverter` embodies `convertToUpper` and when `caseConverter` is invoked in the `changeCase` function body, it will actually invoke `convertToUpper` to do the job, which is to convert the string to uppercase.

Similarly, you can do this:

```
scala> changeCase("IRFAN",convertToLower)
res13: String = irfan
```

In a nutshell, you are passing a function (i.e., either `convertToLower` or `convertToUpper`) as a parameter to a function (i.e., `changeCase`). This is only possible courtesy of Scala's functional programming nature. It's just like passing a value to a variable. In this context, functions are treated in the same way as variables. Hence the maxim: Functions are first class citizens in Scala.

You will see this concept in action quite often when you use functions like `map` and `filter` on Scala collections in upcoming chapters. Those functions accept another function as an argument to perform powerful operations on collections.

Anonymous Functions

So far, you have used functions by defining their names, right? Then you invoke those functions by using their names. While creating functions, you specify the function name after the def keyword.

But Scala provides another feature, called *anonymous functions*. In Python, the equivalent feature is lambda functions. Anonymous functions don't have a name assigned to them. They are nameless. You generally use these functions when you need to pass a function as a parameter to others, like we did before.

Here's the syntax to define anonymous functions in Scala:

```
(parameterName:parameterType)=>function_body
```

In line with the stated syntax, let's create a couple of anonymous functions to strengthen the concept further:

```
(name:String)=>name.ToUpperCase
(name:String)=>name.ToLowerCase
```

To use them, you can assign them to a variable, thanks to the functional programming support in Scala:

```
scala> val convertToUpperAnon=(name:String)=>name.toUpperCase
convertToUpperAnon: String => String = $$Lambda$1137/45854145@5a5183ed
```

Here is what's happening:

- You defined a variable called convertToUpperAnon.

- You assigned a value to that variable, which is everything to the right of the equals sign =.

- In this case, it's an anonymous function that accepts one parameter (the name of type String) and does something to it (converts it to uppercase):

  ```
  (name:String)=>name.toUpperCase
  ```

120

- Now you can invoke this anonymous function as follows:

```
scala> convertToUpperAnon("irfan")
res22: String = IRFAN
```

You used the variable that you used to store the anonymous function. That variable is a container for the function and that function accepts one parameter, so what you include in parentheses is passed as a parameter to that function.

Note that you did not use def in this scenario.

Similarly, while invoking the changeCase function that we defined a while ago:

```
scala> changeCase("irfan",(x:String)=>x.toUpperCase)
res21: String = IRFAN
```

Unlike before, this time you passed an anonymous function ((x:String)=>x.toUpperCase) as a second argument. Remember the function definition of changeCase? It was like this:

```
def changeCase(givenName:String,caseConverter:(String)=>String)
= caseConverter(givenName)
```

As per the function definition, in its second argument, it's expecting a function that can accept a String as a parameter and returns String as the output. Isn't your anonymous function doing the same thing? It accepts a string parameter (x:String) and returns String as the return type (x.toUpperCase). That is why you were able to use this anonymous function in this instance as well.

There's so much to cover about this amazing feature of Scala, but as this book is meant to be an introduction of Scala specifically in the context of Big Data development, I suggest you research this further on your own, leveraging the foundation that I've established here.

In this chapter, you were introduced to the powerful concepts of functions and learned about a number of caveats associated with them. Be sure to practice this concept as much as you can, as these are heavily used in all aspects of programming.

EXERCISES

- Understand the difference between functions and methods in the context of functional programming in Scala.

- Consider a use case of recursion and try applying it.

- Try using functions within functions, i.e., local functions. Understand whether you can refer to inner functions in outer scope.

- Understand whether variables were copied by value or by reference in Scala and what the implications of doing so are.

- Understand the best practices of functions, in that they should be designed so that they perform one and only one task.

CHAPTER 9

Collections

So far, we've been working with variables and data types. We have seen that every line in Scala is an expression that returns a value. You can also group multiple expressions in the form of code block so that the result of the last expression gets returned. But in that case, only one value is returned and the variables we've been working with so far have one value in them.

In many scenarios, you work with data types or data structures that can hold multiple values in them. More specifically, variables of such data structures/types can hold more than one value in them. In Scala, such types are called *collections*. Scala's ecosystem of collections is quite powerful. You will find that this chapter is one of the longest in this book and this is for good reason: If you are able to develop a strong skillset using Scala collections, it will become a lot easier to use Apache Spark APIs. That's because in Spark, you will be dealing with collections as well (although they are distributed ones but the use is highly similar). That's why I've invested a lot of time in this chapter; it's beneficial to everyone.

Real-Life Examples of Collections

To give you more real-life examples of such collections, here are some:

- Collection of grocery items (apples, bread, eggs, butter, and oil)

- Collection of students (Irfan, Raza, Arslan, Ahad, and Hammaad)

© Irfan Elahi 2019
I. Elahi, *Scala Programming for Big Data Analytics*,
https://doi.org/10.1007/978-1-4842-4810-2_9

- Collection of temperature values (100, 98.8, 101, 102, and 95)

- Collection of employees who are still employed at a company (true, true, false, and false) (where true represents an employee who is employed and false means the employee isn't)

What's common in these collections? Apart from the fact that all of them are collections, there is another similarity. The type of data is the same inside the collection, for example, the first and second collections' elements are all of `String` type, the third collection has `Numeric` values, and the fourth has `Boolean` values.

In Scala, when you work with collections like List, Arrays, and Sequence (where List/Array are specific instances of Sequence), it's expected that the data will be of the same type.

Now consider another flavor of collection:

- Customer to number map/dictionary ("Irfan" ➤ 10019181, "Raza" ➤ 1219121)

- Country to capital map/dictionary ("Pakistan" ➤ "Islamabad", "Australia" ➤ "Canberra", "USA" ➤ "Washington")

In these examples, the collections contain items in the "something_first" ➤ "something_second" format. In this context, "something_first" can be thought of as *key* that you can use to look up a *value* (which is "something_second"). It's similar to a phone directory or a contact list on your smartphone, whereby you use names (key) to look up numbers (value).

This concept is quite powerful as it provides an instant indexing feature. If you know the key, you can instantly look up its value; you don't need to traverse the whole collection to find your value. In Scala (and other languages), there is a data structure called Map (also called *hash-map*) that contains such data. Map is a type of Scala collection that we'll explore in detail in this chapter. However, there are some challenges when it comes to building this data structure that are related to collision, whereby keys should be unique. If they are not unique, they can be handled in different ways, such as overwriting the value against that key. To elaborate the point further, if a key called "Irfan" exists in the Map collection with a value 10019181 and you try to insert another key-value pair like "Irfan" ➤ 200, it will overwrite the previous value. These are more data structure-related concepts and I don't want to overwhelm you with these.

Even in the case of Maps, specifically in Scala and other type-safe languages like Java, once you define the type of key and values, they must remain the same. For instance, if your key is String and the value is Number, each key-value pair should follow that same format. You can't mingle different types. However, key and value can be of any type.

There is another flavor of collections in Scala:

- Record of a patient: 1, John Doe, St Mary Hospital, Dr. Robert Jones

- Record of customer: 10, Tony Stark, Pre-paid, Lahore, Pakistan, true

What's the first thing that you notice here? You may have noticed that in each collection, the elements are different types. In the case of the patient record, we have id (1, Int), name (John Doe, String), hospital name (St Mary Hospital, String), and doctor name (Dr. Robert Jones, String).

In the previous examples, all the elements were the same type. In the current example, a collection can have elements of different types. For such scenarios, you use *tuples*. This collection exists in Python as well and it's called the same thing in Scala.

Lastly, there is another type of collection in Scala called *sets* in which order doesn't matter and it doesn't allow duplicate values. If you have studied sets in your elementary mathematics, these exact properties exist in mathematical sets as well and they manifest in the world of Scala (and in Python). So when you create a set, you cannot be certain of the order; if you try to create a set from a collection that has duplicate values, it will remove the duplicates (thus it's a quick way to remove duplicates in your collection).

Another dimension of understanding collections in Scala is related to mutability and immutability. In Scala, you will find that collections fall into two categories:

- Mutable

- Immutable

This means that once you initialize an immutable collection in Scala, you can't add/remove/update elements from it. In the mutable ones, you can. There are still some caveats to it. We cover these later in this chapter.

Now that you have an intuitive understanding of collections, let's move to the hands-on part of it where the actual fun lies.

Understanding Lists

One of the most commonly used collections in Scala is the List collection. For techie folks, it's actually a linked list. Scala lists are optimized by the compiler and it's efficient to use them. They do have certain limitations, such as not being fit for parallel programming.

In the examples in the introduction of this chapter (collection of grocery items, collection of students, and so on), all the elements are of the same type. It's time to create your first list:

```
val myIntegerList = List(1,2,10,30)
val myStringList = List("New York", "Melbourne", "Islamabad",
"Istanbul")
```

You can create these lists in Scala REPL, as shown in Figure 9-1.

```
scala> val myIntegerList = List(1,2,10,30)
myIntegerList: List[Int] = List(1, 2, 10, 30)

scala> val myStringList = List("New York", "Melbourne", "Islamabad", "Istanbul")
myStringList: List[String] = List(New York, Melbourne, Islamabad, Istanbul)
```

Figure 9-1. *Creating lists in Scala*

Using Scala REPL's feature that outputs additional information, here's what we can gather:

- myIntegerList is a list of a specific type called Integer. The Scala compiler looked at the elements of the list and saw that all the elements are of the Integer type and thus it inferred the type.

- Similarly for myStringList, the inferred type is String for the same reasons.

You can have elements of different types in a list:

```
val mixList = List(1,"New York","Melbourne",2,"Islamabad")
```

However, there will be consequences of doing this, which in Scala REPL looks like Figure 9-2.

```
scala> val mixList = List(1,"New York","Melbourne",2,"Islamabad")
mixList: List[Any] = List(1, New York, Melbourne, 2, Islamabad)
```

Figure 9-2. *Creating a mixed list in Scala*

When you create a `mixList`, where elements are of different types like `String` and `Integer`, Scala assigns the highest type (`Any`). Scala provides another collection called `Tuples` that allow you to store different data types and we'll cover that later in this chapter. However, in the previous example, `mixList` is still a list (of type `Any`).

For object oriented programming enthusiasts, something called *polymorphism* should ring bells, right?

Indexing List Elements

So after creating a list, what else can you do? You can access each element. You do that via indexing, using parentheses and specifying a number as follows:

```
scala> myIntegerList(0)
scala> myIntegerList(2)
scala> myStringList(0)
scala> myStringList(3)
```

This is illustrated in Figure 9-3.

```
scala> myIntegerList(0)
res1: Int = 1

scala> myIntegerList(2)
res2: Int = 10

scala> myStringList(0)
res3: String = New York

scala> myStringList(3)
res4: String = Istanbul
```

Figure 9-3. *Accessing lists in Scala*

Recall, as with many other programming languages, the indexing starts at zero. This means that if you want to access the first element, you use 0 and count up from there. Figure 9-4 highlights this idea.

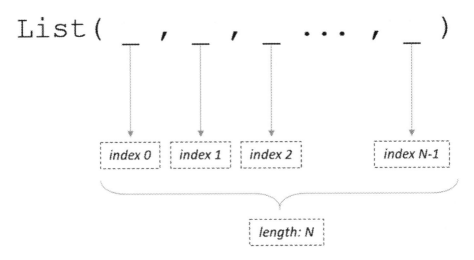

Figure 9-4. *Indexing a list in Scala*

Thus, myIntegerList(0) and myStringList(0) provide the first element of both lists, myIntegerList(2) gives the third element, and myStringList(3) gives the fourth.

What Can You Store in Lists?

You can store any JVM object that you want in lists. By this, I mean that you can store whatever type you want in a list; you aren't restricted to just the primitive types like Integer or String (remember in Scala these types aren't actually treated as primitives, as is done in other languages like Java). So you can create your own type and store it in a list. For example, to represent a person, I create a type via case class as follows:

```
case class Person (name:String, age:Int, employer:String,
isMarried:Boolean)
```

So just think of this expression as a new type that allows you to store data/record about a person, like name, age, employer, and isMarried. Just don't overthink and settle on this understanding. This is another type that you created for yourself.

Previously, you stored integers and strings in a list. We can also create a list of our freshly created Person type, like this:

```
val listOfPersons = List(Person("irfan",30,"Deloitte",true),
Person("Tony Stark",45,"Avengers", false), Person("Neo",34,
"Matrix",true))
```

In Scala REPL, you do something similar to what's shown here:

```
scala> :paste
// Entering paste mode (ctrl-D to finish)

case class Person (name:String, age:Int, employer:String,
isMarried:Boolean)
val listOfPersons = List(Person("irfan",30,"Deloitte",true),
Person("Tony Stark",45,"Avengers",false), Person("Elon Musk",
34, "Tesla", true))

// Exiting paste mode, now interpreting.

defined class Person
listOfPersons: List[Person] = List(Person(irfan,30,Deloitte,
true), Person(Tony Stark,45,Avengers,false), Person(Elon Musk,
34,Tesla,true))

scala> listOfPersons
res0: List[Person] = List(Person(irfan,30,Deloitte,true),
Person(Tony Stark,45,Avengers,false), Person(Elon Musk,
34,Tesla,true))
```

It didn't complain, which means it's a valid operation. Notice the type of List. It's List[Person]. Previously, we had List[Int], List[String].

Thus the point is—you can store whatever object you want in a list.

I could fill books with what you can do with lists, but it's beyond the scope of this book. I just cover some of the functions and ways you can operate on lists.

Widely Used List Operations

In this section, we look at the common and widely used list operations.

List Size

You can determine the length of the list via the .length field as follows:

```
myIntegerList.length
```

It will return an integer, indicating the length of the list.

Note   A *field* is an object oriented programming related concept, but as of now, you can think of a field as something that represents the properties of objects. Thus, in the case of List, if you consider it an object, the length field is a property of it, indicating the length of it.

Basic Statistics of Lists

You can find the min and max elements of a list by using the .min and .max fields in a list. Give it a shot by creating a list of integers and finding the min and max of the list.

Converting a List to a String

Many times you will need to create a string from the elements of a list. To do that, you use the .mkString method. It also allows you to specify the separator. See Figure 9-5.

```
scala> myStringList.mkString(";")
res14: String = New York;Melbourne;Islamabad;Istanbul
```

Figure 9-5. *Creating a string from a list*

As you can see in Figure 9-5, the returned value is a string. It concatenated/joined each element of the list and added the delimiter that you specified. Try using mkString without a delimiter and see what happens.

Iterating Over Lists

Owing to the functional programming characteristics of Scala, there are a number of cool ways to traverse the elements of a list. The following are commonly used:

- foreach

- map

- Loops (covered in the next chapter)

First, don't confuse this map with the hash-map I highlighted before. That was a data type. This is a function that allows you to traverse elements on the list.

Using the map Function to Iterate Over Lists

Let's look at the map function. It's one of my favorites. A similar function exists when you do programming using Spark APIs. In fact, many of the operations that you do on a list will be similar to what you do with Spark APIs. Thus, you're developing a sound foundation of Scala in order to be a Big Data developer.

The best way to understand map is, as usual, by using it.

Unlike other methods that you used—such as split or replace—on the String data type, map accepts functions as arguments, as discussed in Chapter 8.

Getting to Know Functional Programming Concepts

Functional programming is a diverse topic, but as elaborated previously, what you need to know is that if a language supports functional programming constructs, then functions become first-class citizens in that language. That means that you treat functions the same way you treat other data objects. You can assign them to variables or pass them as parameters to functions. You can pass a function as a parameter to another function. This concept is foundational to understanding the upcoming two functions (map and foreach). These two, in some shape and form, also exist when you use Apache Spark APIs.

One thing that you may understand by now is that the map function accepts a parameter and that parameter is a function:

```
map(f(x))
```

where f(x) is a function that gets passed to map.

Here is another important property of the operational characteristics of the map function: the function that you pass to it as a parameter will be applied to every element in the list. So f(x), no matter what it does, will be applied one-by-one to each element in the list. For each element in the list, the function will be called and that function will do something on that element and will return a value.

Figure 9-6 illustrates this idea.

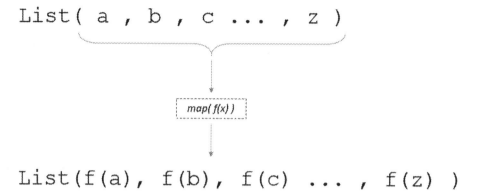

Figure 9-6. *Highlighting how the map function works on a list*

Your list will be transformed by that function; each element of the list will be transformed by that function.

Let's dissect this idea with an example. Let's say that we define a function that creates the square of each integer.

```
def squareThis(givenParam:Int):Double = math.pow(givenParam,2)
```

So you've defined this function and you can call it independently. When fed an integer number, it returns the square of that number. Simple.

Now let's define a list:

```
val numberList = List(1,3,5,7,9)
```

This list contains some numbers. Nothing fancy.

Now what if you want to get the square of each element of this list? You want to apply the squareThis function to each element of this list. In non-functional programming languages, you typically would write a loop that iterated through the elements and did something. In a functional programming language like Scala, you can express the same task in a more expressive way by using the map function:

```
numberList.map(x=>squareThis(x))
```

Wait, what? You may be wondering what (x=>squareThis(x)) means. You might be able to identify the part using the squareThis function that we used before. But, what's this new notation?

There is a lot of technical reasoning behind this, but a simple way to understand is that in the expression

```
(x=>SquareThis(x))
```

The x before => represents a parameter. In this case, it represents each element of the list on which you are calling the squareThis function. So, in the first iteration, x will have value of 1, and then 3, then 5, and so on. So x will hold values of the list, one by one, on which the map function is invoked.

The next part is easy; the right side of => represents the function that you want to call. So you call squareThis and pass it x, which in each iteration represents each element of this list.

Scala provides a simpler way to invoke such functions:

```
numberList.map(squareThis)
```

This is equivalent to what you wrote before. This is generally used when your function accepts one parameter, as is applicable in this context.

Just like you defined a function to be used in the map function, you can invoke the functions that come with data types. For example, if you are operating on a String data type as follows:

```
val stringList = List("Australia","USA","UK","Malaysia","Singapore")
```

And you want to find the length of each element in the list, you can simply use this approach:

```
stringList.map(x=>x.length)
```

It will give you the length of each element of the list.

EXERCISES: LISTS AND THE MAP FUNCTION

There are many ways you can use the amazing map function. Here are some example exercises to strengthen your grip on this function:

- Create a list of numbers and return `true` if an element is even; otherwise, return `false`.

- Create a list of strings and extract the first and last character of each string.

- Load a file in Scala and load its content in a list. Then iterate through each line, one by one.

What Is Returned When You Use the Map Function?

There's an important point to be noted about map. Using the function that you pass, it applies that function to each element of the list and then returns a list. You will find that this behavior is quite different from other functions, like `foreach`, which actually don't return anything.

In the case of map, if you invoke it on a list, you get a list. However, here is another important caveat. It's not necessary that the list be of the same type. You've seen this already, when you created a list of strings and found the length of each element, the type of list that was returned was different.

Do that exercise again and notice that the type of your original list (`List[String]`) and the type of the list that's generated by invoking map with that specific `.length` function in it. I know that `.length` isn't a function, but for the sake of brevity, let's proceed with this understanding. So, the list type will be `List[Integer]`. You converted the list from one type to another, as further indicated in Figure 9-7.

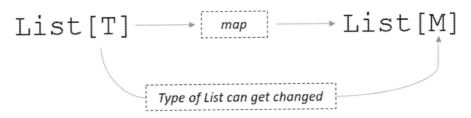

Figure 9-7. *Return type of map function on list*

You will do that a lot in Spark programming, as you will be converting RDD (a fancy name for distributed collection of data that's loaded in Spark and acts like a Scala list) from one form to another.

Using foreach on Lists

Now with this understanding of map, let's look at foreach. It is pretty simple. It operates the same way as map in that you pass a function that you want to apply to each element of the list. The only caveat is that in foreach, the return type is Unit. It returns no value. Unlike with map, in which you get a list, albeit a changed one, with foreach, you don't get a list in return.

Figure 9-8 illustrates this further.

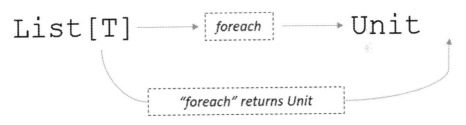

Figure 9-8. *Return type of foreach function on List*

The notion of foreach is that you use it to do some processing on each element. Typically, it's used to print elements of a list. More technically advanced use cases of foreach include when you want to interact with a third-party system. For example, if you want to store each element of a list to a database. In such use cases, you don't need the list to be returned. Rather, you use the elements of the list to interact with other sub-systems. This is quite a powerful concept and is heavily used in Apache Spark to interact with other systems like databases and message queues.

For now, try using foreach as follows:

```
numberList.foreach(x=>squareThis(x))
```

When you execute it, you will see blank output. Why? Because it returns no value and thus nothing gets printed.

Try this:

```
numberList.foreach(x=>println(s"My id is $x"))
```

In this case, you will see some output on the screen courtesy of the println statement. So println will print each element onscreen. But it doesn't mean that foreach returns a value. It will still return nothing. Want proof? Try assigning the following to a variable and view the type:

```
numberList.foreach(x=>println(s"My id is $x"))
```

It will be Unit.

Using the filter Function on Lists

The filter function is heavily used with lists and has a cousin in the world of Apache Spark APIs. filter is used to select certain elements of a list that satisfy a particular condition and then return a list.

There are some similarities between the filter and map functions, as follows:

- Like map, filter also returns a list. That's why it's different from the foreach function.

- Like map (and foreach), it expects a parameter in the form of a function.

- Like map (and foreach), it applies the function that you pass as a parameter to each element of the list.

Now with the similarities highlighted, let's discuss how filter is different:

- The function that you pass to the list must return a Boolean value (true or false). Generally, the function that you pass performs some conditional checks on each value of the list and then returns either true or false.

- If a function for a specific value of a list returns true, that value gets retained in the final list. If a function for a specific value of a list returns false, that value is discarded and doesn't appear in the final list.

Let's understand this with an example, shall we?

Consider an example in which you have a list of numbers (from 1 to 10) and you want to select only the even numbers from this list. Here's how you can do that via the filter function.

First, we define a function that we will pass to filter as a parameter:

```
def isEven(givenParam:Int):Boolean = givenParam%2 == 0
```

```
scala> val numberList=(1 to 10).toList //this is yet another
method you can quickly create a list. This will create a list
consisting of numbers from 1 to 10.
numberList: List[Int] = List(1, 2, 3, 4, 5, 6, 7, 8, 9, 10)
```

```
//if we use .filter method and pass that isEven function, in
the same way we did for map, we get:
```

```
scala> numberList.filter(x=>isEven(x))
res2: List[Int] = List(2, 4, 6, 8, 10)
```

As you can see, only the even numbers exist in this list. This is because only these list elements returned true per the function that we passed in the filter method.

Another important point to understand is that as a result of using filter:

- You get a list that has the same type as the original list.

- The number of elements in a list returned by filter can be the same or less than the original list.

Thus, if you have a use case where you want to filter/select/exclude/include specific elements of a list, you should use filter, not map or foreach.

Using the Reduce Operation on Lists

So far you have seen functions that operate on each element in a list. If you want to perform some aggregation on the elements in a list, you can use the reduce operation on them. The reduce operation works by taking all the elements in a list and aggregating them in some way to produce a single value. Specifically, it uses a binary operation to combine the first two elements of the list. Once it's done, it uses the result of that first aggregation and proceeds to combine that with the next element in the list. This process continues until the end of the list.

One point to note is that reduce works when you use a binary operator (an operation that uses two operands, such as addition, multiplication, or even string concatenation).

Let's say you have a list of numbers and you want to find the sum of the elements of the list. You can probably use reduce (in addition to using the .sum function):

```
scala> val list=(1 to 10).toList
list: List[Int] = List(1, 2, 3, 4, 5, 6, 7, 8, 9, 10)

scala> list.reduce((x,y)=>x+y)
res11: Int = 55

scala> list.reduce(_+_)
res12: Int = 55
```

Let's look at the previous code snippet:

- In the first step, you created a list by calling .toList on the (1 to 10) range collection.

- Then you used the reduce function on that. In the reduce function, you specified parameters like ((x,y)=>x+y), which means that when it encounters two elements of a list, it will apply the operation of addition on them and will aggregate them.

Under the hood, the x variable acts like an accumulator, which gets initialized to zero (because we are using addition; if it's multiplication, it will be 1). It starts adding elements of the list one by one. Initially, the value of x will be zero and then it will add the first element to it. In the second iteration, x will add the second element of the list (2) and it will become 3, and so on. So it will act like an accumulator. Figure 9-9 illustrates this idea.

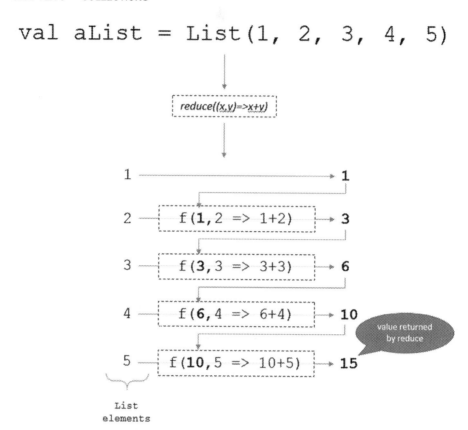

Figure 9-9. Reduce operation on a list

Scala provides another syntactical way to represent ((x,y)=>x+y) as (_+_) which means the same thing. It's a concise way to write the same thing.

List Equality Check

You can quickly check if two lists are the same (have the same elements in them) by using the == operator. It's a quick way to check list equality and it is somewhat different from other languages.

Alternative Ways to Create Lists

You've already seen two ways to create list. Scala provides another way to create lists, specifically via the con (::) operator and Nil, as follows:

```
scala> val aList = "a" :: "b" :: "c" :: Nil
aList: List[String] = List(a, b, c)
```

Specifically, Nil represents an empty list or lists of zero length.

EXERCISES: LISTS

- Explore how to append two lists, append an element to a list, and take a union or intersection of two lists on your own. When you do these operations, does it change/mutate the original list or does it return a new instance of list?

- Create a list of strings and then try using reduce to achieve the same output as you would get by using the .mkString function.

Creating Sets

As mentioned earlier, the elements in a set should be unique. There is no notion of order in sets.

Here's how you can create sets in Scala:

```
scala> val aSet = Set(1,10,121)
aSet: scala.collection.immutable.Set[Int] = Set(1, 10, 121)
```

You must use elements of the same types in Set.

If you try to create a set with duplicate elements, it will simply discard the duplicate values:

```
scala> val aSet = Set(1,10,121,10)
aSet: scala.collection.immutable.Set[Int] = Set(1, 10, 121)
```

I tried creating a set with the value 10 duplicated. It created a set with that duplicate value removed. Whenever I need to remove duplicates from a list, I convert my list to a set, which automatically removes the duplicates. Consider this example:

```
scala> val duplicateList = List(1,10,121,10)
duplicateList: List[Int] = List(1, 10, 121, 10)

scala> duplicateList.toSet
res24: scala.collection.immutable.Set[Int] = Set(1, 10, 121)
```

If you try to access an element of a set like you do with a list, you will find surprises. If you use parentheses with sets, as follows:

```
aSet(N)
```

It will not fetch the element at the N position. Rather, it will check the membership of that element; it will tell you whether that element exists in the set.

```
scala> aSet
res25: scala.collection.immutable.Set[Int] = Set(1, 10, 121)

scala> aSet(1)
res26: Boolean = true

scala> aSet(10)
res27: Boolean = true

scala> aSet(20)
res28: Boolean = false
```

As sets are not ordered collections, accessing the first or second element doesn't make sense. That's why aSet() simply checks whether an element exists.

You can traverse elements of a set like you do in lists, by using map or foreach.

Understanding Map Collections

We've spent a great deal of time on lists, which is important as they are foundational and widely used in Scala. Now let's look at the Map collection. First things first, this Map collection has no association with the map function that you used before in lists that allowed you to transform lists. Map, in this context, is a type of Scala collection and can hold a collection of values. It's like a list but with a twist.

Remember the example that I gave before about the different types of collections we encounter in real life? I specifically highlighted a scenario where you have constructs like key-value pairs and you use a key to look up a value. You use the contacts in your smartphone to look up a number via your contact name. The Map collection works in the same way.

In the case of the Map collection, you have a key-value pair. Each element of a map consists of this pair—on position 0 (which is the first position in Scala collections), there is a key-value pair. That key and value will not occupy different index positions in Map. They will exist at the same index position. Comparing this with list, you have one object/element at each index. Now in Map, you have a pair of key-value. Interesting!

Figure 9-10 illustrates the concept further.

```
Map(Key->Value, Key->Value ... Key->Value)
```

Figure 9-10. *What a Map collection consists of in Scala*

Let's see how you can create one:

```
scala> val contactsMap = Map("thor"->918101,"captain america"-
>1281281,"hulk"->91921921)
contactsMap: scala.collection.immutable.Map[String,Int] =
Map(thor -> 918101, captain america -> 1281281, hulk ->
91921921)
```

You created a Map in the previous example that represents contact details for the Marvel superheroes.

Here are some observations:

- Each element in this collection is a key-value pair, represented as "key" ➤ "value".

- In this example, key is of type String and value is of type Int. This type should remain consistent in a map. If you mix and match types, it will pivot to a bigger data type, like Any, like we've seen before.

- Thanks to Scala REPL, the type of this Map is quite clearly highlighted. It's scala.collection.immutable. Map[String,Int], which means two things:

  - It's an immutable map (we will see that there can be a mutable map as well).

  - The type of this map is Map[String,Int], which is based on the fact that the keys in your map are String and the values are all Integer. Scala inferred the type based on the elements/key-value pairs that you specified in the map.

Indexing a Map

If you try to index a map like you do a list—that is, by using (N)—you will be greeted with an error, as shown in Figure 9-11.

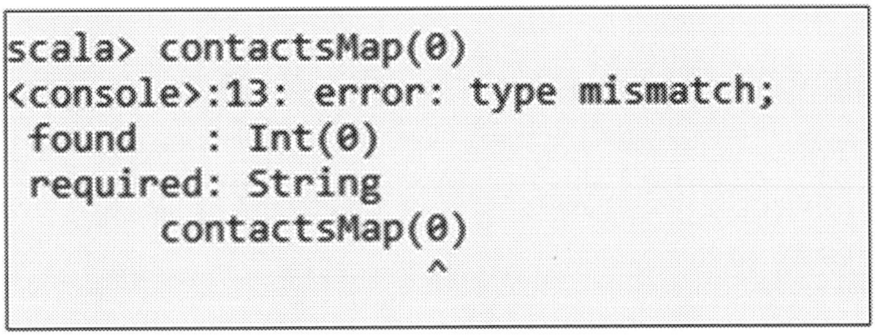

```
scala> contactsMap(0)
<console>:13: error: type mismatch;
 found   : Int(0)
 required: String
       contactsMap(0)
              ^
```

Figure 9-11. *The wrong way to index a map*

The old ways of indexing don't work here; you can't use numerical indexing or positional indexing like you can with lists.

So how do you index an element? You do so by passing the key in parentheses. If that key exists in your map, its value will be returned, as shown in Figure 9-12.

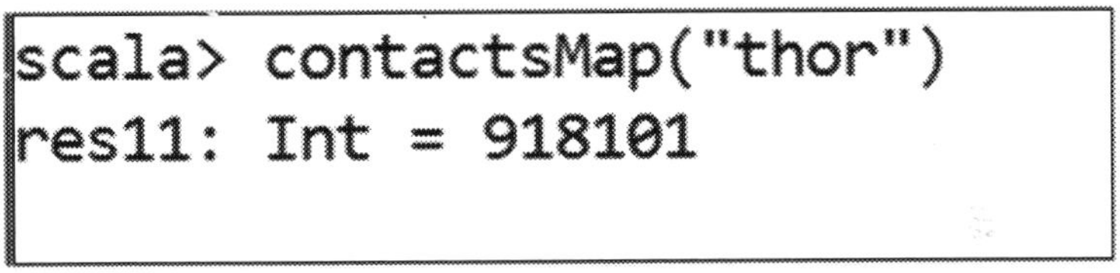

```
scala> contactsMap("thor")
res11: Int = 918101
```

Figure 9-12. *Indexing a map properly*

You looked up the contact number of Thor by using his name as the key. It's the same behavior of looking up a contact number in your mobile phone via contact name.

So what if you pass a key that doesn't exist in your map? You will get error, as shown in Figure 9-13.

```
scala> contactsMap("iron man")
java.util.NoSuchElementException: key not found: iron man
  at scala.collection.immutable.Map$Map3.apply(Map.scala:167)
  ... 28 elided
```

Figure 9-13. *Indexing a map using a key that doesn't exist*

Rightly so. A safer and cleaner way to do lookups in a Scala map are as follows:

```
scala> contactsMap.getOrElse("iron man","not found")
res14: Any = not found
```

This involves using .getOrElse. This code says that if a key is present, return its value; otherwise, return the value that you specified as the second argument ("not found") in this example.

Uniqueness of Keys in Maps

Another caveat about Map is that its keys have to be unique. What will happen if you specify keys that are the same? Go on and give it a try. It won't give you an error. But do see what happens.

```
scala> val contactsMap = Map("thor"->918101,"captain america"-
>1281281,"hulk"->91921921,"thor"->99021)
contactsMap: scala.collection.immutable.Map[String,Int] =
Map(thor -> 99021, captain america -> 1281281, hulk -> 91921921)

scala> contactsMap("thor")
res0: Int = 99021
```

The key that you specified as the last will be used and the previously defined key-value pair for the same key will be ignored. If you look up based on a key and if that key exists more than once, which value should be returned? Scala's heuristic is to consider the latest key-value pair when there are duplicate keys. As in the previous example, when you used contactsMap("thor") to look up the value against Thor, it returned the latest value in the map.

However, there is no such restriction on values. They don't have to be unique. This makes sense.

Alternative Ways to Create Map Collections

Like with lists, Scala provides another way to create Map collections as well. With this alternate method, you basically make two changes compared to the previous method:

- You surround your key-value pair with parentheses

- You replace -> with a comma (,)

Here's an example:

```
scala> val contactsMap = Map(("thor",918101),("captain america",
1281281),("hulk",91921921),("thor",99021))
contactsMap: scala.collection.immutable.Map[String,Int] =
Map(thor -> 99021, captain america -> 1281281, hulk -> 91921921)
```

There is no difference in the method you choose to create Scala Map collections.

Manipulating Maps

If you want to see which keys are available in your Scala Map, just use the .keys attribute:

```
scala> contactsMap.keys
res1: Iterable[String] = Set(thor, captain america, hulk)
```

Similarly, if you want to get values out of the Scala Map, use the .values attribute:

```
scala> contactsMap.values
res2: Iterable[Int] = MapLike.DefaultValuesIterable(99021,
1281281, 91921921)
```

If you want to see if a particular key exists in your Scala Map, you can use the contains method, which will return true if the key is present and false if not:

```
scala> contactsMap.contains("thor")
res3: Boolean = true

scala> contactsMap.contains("black panther")
res4: Boolean = false
```

If you have the urge to assign/update a value against a particular key, you will be discouraged because this is an immutable Scala map.

Let's say Hulk's contact number changed and you want to update it in your current Scala Map instance:

```
scala> contactsMap("hulk") = 81729191
<console>:13: error: value update is not a member of scala.
collection.immutable.Map[String,Int]
       contactsMap("hulk") = 81729191
```

You will get an error. For such luxuries, you have to use mutable collections, which we discuss in coming sections. Also, even if you use mutable collections, you still have to be cautious of the type of value that you are trying to update.

Iterating Through Maps in Functional Style

Map also gives you a map function. Using that map function, you can transform your Map collections. However, since you are dealing with a key-value pair, things will be slightly different.

For example, say you need to make each of your keys (contact names in this example) uppercase.

To do that, you have to transform each element and get an object/map out of the results. Therefore, using the map function makes sense (instead of foreach, which would return Unit or filter, which is better suited for excluding/including certain elements).

Thus, we use the following code:

```scala
scala> contactsMap.map{case(x,y)=>x.toUpperCase -> y}
res24: scala.collection.immutable.Map[String,Int] = Map(THOR ->
99021, CAPTAIN AMERICA -> 1281281, HULK -> 91921921)
```

Here is what's happening here:

- Like before, you use a map function, in the same way you used it in lists.

- In the body of the map function, you use brackets {} instead of parentheses. Otherwise, you would get an 'illegal start of simple expression' error.

- You use case(x,y) where you typically represent parameters. You are basically mapping keys to x and values to y. Remember in lists like this one that x represented the element in each iteration:

  ```scala
  myList.map(x=>f(x))
  ```

  Similarly, in each iteration, (x,y) will represent the key-value pair and you use that via case(x,y).

- You operate as usual to make whatever transformations you want on the right side of =>. In this case, you use x.toUpperCase to convert keys to uppercase and you keep y as it is.

This can appear to be complex at first, but if you practice it a bit, it will make sense. There is another way to use the map function in this example, but I'll highlight that once I cover the Scala tuples. On a similar note, you can also use filter and foreach.

Understanding Tuples

So far, we covered the List and Map collections. In those collections, recall that the data types used in the elements were the same type. You created a map where keys were of a specific data type and the values were a specific data type. If you try to mix and match different data types, Scala assigns a higher data type, like Any, which in some scenarios can lead to ambiguities.

Referring back to the real-life examples of collections mentioned at the start of this chapter, specifically the example related to the nature of rows present in a spreadsheet or table in a database system. In that scenario, collections may have different data types. For example, if you store details about a customer like so:

```
id | name | phone number | location | isActive
1  | Irfan| 919191       | Pakistan | true
```

You can see that you are dealing with different data types. Clearly, previous collections may not be of much help here. That's where the Tuples collection comes into play. It's also found in other languages like Python. The Tuples collection addresses the pain point that we just observed—it can store data of different types. Coolness.

Let's start exploring tuples in Scala by creating one:

```
scala> val aTuple = (1,"customer1","australia","prepaid",true)
aTuple: (Int, String, String, String, Boolean) = (1,customer1,
australia,prepaid,true)
```

We created a tuple in this example. Here are some points:

- To create a tuple, you just use a comma-separated list of values that you want to be included in the `Tuple` collection and surround them with parentheses.

- You can see that this tuple also has a type (`Int, String, String, String, Boolean`), which represents the corresponding elements in your tuple. As you can see, we used a variety of data types in our tuple.

Indexing Tuples

What do we do when we create collections? We access the elements!

Unlike `List` and `Map`, there are two differences when it comes to indexing elements. First, we don't use parentheses to index the elements of a tuple. Rather, we use this different syntax:

```
aTuple._N
```

where `N` stands for the position of the element that you want to access.

Secondly, in tuples, the index starts at 1! Shocking maybe, but that's how it is. So to access the first element of your tuple, you will use this:

```
aTuple._1
```

Instead of this:

```
aTuple._0
```

This concept is illustrated in Figure 9-14.

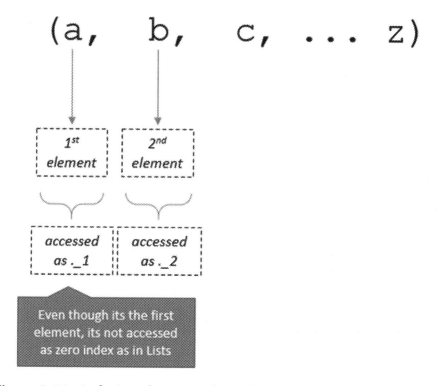

Figure 9-14. *Indexing elements of a tuple*

Thus, to access the first element of the tuple that you just created, you use:

```
scala> aTuple._1
res0: Int = 1
```

To access the second, you use:

```
scala> aTuple._2
res1: String = customer1
```

And so on.

Iterating Over Tuples

Unlike List and other collections like Map, you will find that there are no map, foreach, and filter methods available with tuples.

So how do you iterate through a tuple? It's done via productIterator.

When you use productIterator on a tuple, it returns an Iterable, which is a trait in Scala. Now don't worry if you don't understand what traits are (they are equivalent to interfaces in Java for Java folks); traits are like types in a way and define certain characteristics, which are then further implemented by sub-types. For now, just think of Iterable as one type of collection like a list.

So if you use productIterator, you get Iterable and you can convert an Iterable to a List, a type with which you are quite comfortable by now, right? So how do you do this? Here's how:

```scala
scala> aTuple.productIterator.toList.foreach(println)
1
customer1
australia
prepaid
true
```

So here's what happened in this code snippet:

- When you use productIterator on aTuple, you get Iterable (you can verify that by calling aTuple.productIterator).

- You then convert Iterable to a list via the toList method.

- Once it's a list, you use any of the methods that List supports, like foreach, to do whatever you want.

So you've seen one of the ways to convert a collection from one type to another by using `toList`. Keep a note of that as we may use that again.

Now, to find size of a tuple, you can use `productArity` like so:

```
scala> aTuple.productArity
res7: Int = 5
```

Try experimenting with other functions/attributes available with tuple, like `toString`, and research them.

Note   If you are using Scala 2.10 or before, you can't have more than 22 elements in a tuple. This limitation was addressed in subsequent versions of Scala.

Alternative Ways to Create Tuples

There are other ways to create tuples as well. Here are some:

```
scala> val twoElementTuple=Tuple2("irfan","elahi")
twoElementTuple: (String, String) = (irfan,elahi)
```

You use `TupleN`, and then pass `N` elements in parentheses to create a tuple of size `N`. It's mostly a matter of preference when it comes to choosing which syntactical style to use for creating tuples.

Understanding Mutable Collections

So far, you have noticed that Scala emphasizes immutability quite a bit. You've been creating immutable variables so far. You've also been using immutable collections so far. But again, there can be scenarios where you need mutable collections. For example, you want to add elements to a list and receive input from a user, or you want to add key-value pairs to a map to keep track of the frequency of words in a file, whereby each word is a key and each word's value is the number of times it appears. Scala doesn't leave you in the lurch when it comes to this. Let's look at a handful of mutable collections.

One of the most commonly used mutable collection is ListBuffer. Think of it as a mutable variant of list.

Here are some examples of using ListBuffer:

```
scala> import scala.collection.mutable.ListBuffer
import scala.collection.mutable.ListBuffer

scala> val aMutableList = ListBuffer(1,10,91,121)
aMutableList: scala.collection.mutable.ListBuffer[Int] =
ListBuffer(1, 10, 91, 121)

scala> aMutableList += 500
res29: aMutableList.type = ListBuffer(1, 10, 91, 121, 500)

scala> aMutableList
res30: scala.collection.mutable.ListBuffer[Int] = ListBuffer
(1, 10, 91, 121, 500)
```

Here is a step-by-step explanation of what you did in this code:

- You imported ListBuffer in your namespace.

- You then created a ListBuffer like you create a list.

- Once it's created, to exploit its mutable nature, you added or inserted an element using the += operator.

You can update an element at a particular position as well:

```
scala> aMutableList(0) = -91

scala> aMutableList
res32: scala.collection.mutable.ListBuffer[Int] = ListBuffer
(-91, 10, 91, 121, 500)
```

To remove any element from a `ListBuffer`, use this:

```
scala> aMutableList -=500
res34: aMutableList.type = ListBuffer(-91, 10, 91, 121)

scala> aMutableList -= 10
res35: aMutableList.type = ListBuffer(-91, 91, 121)

scala> aMutableList
res36: scala.collection.mutable.ListBuffer[Int] =
ListBuffer(-91, 91, 121)
```

There is a lot you can do with `ListBuffers`, like traversing through elements using `map` and filtering elements using `filter`. Explore these functions, which will be quite similar to lists.

Implications Related to Mutable Collections

There are two important caveats to mutable collections that we should consider:

- You initialized the mutable collection (`ListBuffer`) as `val`, which stands for immutability, but you were able to manipulate its elements with no issues. So `val` in this context has no power. You can mutate elements of a mutable collection.

- If you try to change the assignment of objects to the `val` variable, it will not allow you to do so. Remember, when you create a variable (with `val`` or `var'`), you actually store a reference to the memory location (called a *pointer*) where the values are stored (whether it's a single variable or a collection). So this memory reference, AKA pointer, when created via `val` will remain immutable even if you use mutable collections.

Let's create another ListBuffer like so:

```
scala> val secondMutableList = ListBuffer(818181, 912121)
secondMutableList: scala.collection.mutable.ListBuffer[Int] =
ListBuffer(818181, 912121)
```

Try assigning this:

```
aMutableList = secondMutableList
```

It won't work, because you are assigning a different memory reference/pointer to the initial variable created with val.

However, if you do the same thing with var, you get this level of mutability as well. Go on and give it a shot.

EXERCISE: LISTBUFFERS

- Try using ArrayBuffer as a mutable collection. Make sure you understand the difference between ListBuffer and ArrayBuffer.

Mutable Maps

Just like we have ListBuffer and ArrayBuffer, we have mutable versions of Scala maps as well. They provide the same characteristics:

- You can add/update/delete the elements of the map (i.e., key-value pairs).

- You can't mutate/change the assignment of reference/pointer if you use val.

So let's try using them:

```
//first you import the mutable library into your name-space.
It's suggested to not import scala.collection.mutable.Map as
it will be vague as to whether you want to use mutable or
immutable map in your program.
```

```
scala> import collection.mutable
import scala.collection.mutable
```

```
// creating a mutable map is similar to immutable one:
```

```
scala> val mutableMap = mutable.Map("CEO" -> "John Doe", "CTO"
-> "Tony Stark", "Team Lead" -> "Dwayne Johnson")
mutableMap: scala.collection.mutable.Map[String,String] =
Map(CTO -> Tony Stark, Team Lead -> Dwayne Johnson, CEO ->
John Doe)
```

```
//if you want to update a key-value pair i.e. specifically a
value against a key:
```

```
scala> mutableMap("CEO") = "John Cena"
```

```
//if you want to add a new key-value pair in your mutable map:
scala> mutableMap += "Developer" -> "Nate Silver"
res2: mutableMap.type = Map(CTO -> Tony Stark, Team Lead ->
Dwayne Johnson, CEO -> John Cena, Developer -> Nate Silver)
```

Mutable Map gives you the convenience of doing such operations with ease. I suggest exploring other aspects of mutable Maps, like deleting a key-value pair, iterating through the elements, and so on, to develop a deeper understanding of this.

Using Nested Collections

The last topic that I want to explore before I move on is the nested list. It's quite important and heavily used.

So far, you've created a list of integers and strings. Similarly, if you have been following along, you must have worked on the example where you created a list of the Person case class. What about a list of lists? Or lists of maps? Or a map where the value is a list? Or a list of tuples? Is your mind blown? It shouldn't be, because like Int, String, or case class, collections are just another object in Scala. So even though the notion of nested lists may seem daunting at first, if you look at them under the same lens that these are just lists with Scala objects, things will be much clearer.

As always, let's learn by doing. Let's create a list where each element itself is a list:

```
scala> val nestedList = List(
    List("Australia","Pakistan","Malaysia"),
    List("Asia","Africa","Antarctica","Australia","Europe",
    "North America","South America"),
    List("Microsoft","Apple","Facebook","Twitter","Cisco",
    "Netflix","Uber")
    )
nestedList: List[List[String]] = List(List(Australia, Pakistan,
Malaysia), List(Asia, Africa, Antarctica, Australia, Europe,
North America, South America), List(Microsoft, Apple, Facebook,
Twitter, Cisco, Netflix, Uber))
```

Here's what you just did:

- You created a list and assigned it to a variable called nestedList. That's a start.

- Unlike with the previous examples, each element now
  is a list. Specifically, it's a list of Strings. How do you
  access a specific element in a list? Via indexing, right?
  So go ahead and access the first element as follows:

```
scala> nestedList(0)
res3: List[String] = List(Australia, Pakistan, Malaysia)
```

Sure enough, you get the first element, which is a list itself that you specified while creating the list. Now as this element is itself a list, you can index it further. Nothing is stopping you from doing so:

```
scala> val firstElement = nestedList(0)
firstElement: List[String] = List(Australia, Pakistan,
Malaysia)
```

```
scala> firstElement(0)
res4: String = Australia
```

You store the first element of your list in a variable, which is a list, and then you access its first element, which is "Australia" (a String).

Instead of using this intermediate variable firstElement, you can do the same as:

```
nestedList(0)(0)
```

Where the first (0) allows you to access the first element of the outer or main list and then the second (0) allows you to index the inner list.

If you follow these guidelines, working with nested lists won't be hard. You will be working with such structures a lot, especially if you use Apache Spark.

Let's look at another example. Say you created a list of tuples to store information about the owners of famous websites, like this:

```
scala> val nestedListOfTuples = List((1,"irfan","irfanelahi.com"),
(2,"nate silver","fivethirtyeight.com/"),(3,"Mark Zuckerberg",
"facebook.com"))
nestedListOfTuples: List[(Int, String, String)] = List((1,irfan,
irfanelahi.com), (2,nate silver,fivethirtyeight.com/),
(3,Mark Zuckerberg,facebook.com))
```

Unlike before, this is a list of tuples. The previous example was a list of lists. In here, each element is a tuple and each tuple consists of three elements.

Now let's traverse this list and print the owner names as follows:

```
scala> nestedListOfTuples.foreach(x=>println(s"owner name is:
${x._2}"))
owner name is: irfan
owner name is: nate silver
owner name is: Mark Zuckerberg
```

Here's what you did:

- You created a list of tuples.

- You used the .foreach method, which allows you to operate on each element of a list.

- In the body of foreach, x stands for each element of the list. What is each element of your list? Tuple, right? That's why you used x._2 to access the second element of the tuple. Because you know that in each iteration, x will store the tuple. Thus you used the corresponding syntax to interact with it.

I can't emphasize enough how important this concept is, specifically if you intend to use Apache Spark. In Apache Spark, you will often load data from a Hadoop distributed filesystem, operate on it like converting to lists, and then converting to tuples, and so on. So spend some energy understanding this concept. Trust me, it will pay off.

With this, let's conclude this chapter, which seems to be one of the longest chapters of this book! It rightfully deserves to be, as this chapter is the cornerstone of many concepts that you will employ in Apache Spark programming.

ADDITIONAL EXERCISES

- Understand what `Array` is in Scala. How is it different from the other collections that you studied?

- Understand what `vector` is in Scala and how is it different from the other collections that you studied.

CHAPTER 10

Loops

Many tasks in our life require some degree of iteration or repetition. For instance, books have chapters and you start with one chapter and go until the end. If we want to add numbers from 1 to 100, we have to start at 1 and add them together until we reach the last number. If you are logging into an online web application, you may be prompted for the correct combination of username and password until you get it right. The list of such tasks can go on and on, but the gist of the matter is that if you have to model such tasks in programming languages, you will have to rely on specific constructs to do so.

In your journey to learn Scala so far, you have worked with Scala collections and saw various types. You also observed how to traverse through a Scala collection, like lists via the map or foreach function. You can achieve many real-world examples by traversing a collection in Scala. In addition to those higher-order functions like map and foreach, there is another way that you can traverse through a collection—via loop expressions. If you want to execute certain statements multiple times until a particular condition is met (e.g., until the user provides username and password correctly), you can also rely on loops to implement them.

When you use Apache Spark, you generally don't use these loops to process data distributed over a cluster of machines. Rather, you use Apache Spark APIs to traverse/process that dataset. However, in such a pipeline, after Apache Spark processes large scale datasets in a distributed manner, it can return summarized results (e.g., the count/frequency of

© Irfan Elahi 2019
I. Elahi, *Scala Programming for Big Data Analytics*,
https://doi.org/10.1007/978-1-4842-4810-2_10

how many times a particular term appeared in a huge collection of files). You then use loops to traverse those summarized results from Apache Spark and that's where loops usually come in handy.

Types of Loops in Scala

In Scala, there are two main types of loops:

- for loop
- while loop

Let's look at an example of each of these types in order to further our understanding of this topic.

The for Loop

It always helps to get to know the syntax of an expression first. The syntax of the for loop, in its simplest form, looks something like this:

```
for (counterVariable <- Collection) {
  Loop body
}
```

So a concrete example could be:

```
scala> val listOfNumbers = List(1,20,300,-12,121)
listOfNumbers: List[Int] = List(1, 20, 300, -12, 121)

scala> for (i<-listOfNumbers){
     println(i)
     }
1
20
300
-12
121
```

166

In this code example, here is what's happening:

1. We use the for loop by using a for keyword.

2. In (), we specify two things: a variable to hold each value of the list for each iteration and the collection to be traversed.

3. Then there is the loop's body in which we specify which statements we want to execute. This body is surrounded by {}. You can also refer to the variable (i.e., i in this case) in your loop's body for data manipulation.

In the previous code example, you initialized a list consisting of five Integer values. To traverse this list, you used a for loop. In the for loop, you used the variable i to iterate over the list. This loop will execute five times and in the first iteration, the variable i will store the value 1 from the list. With that value stored in the variable, the body of the loop will execute. (It consists of the println statement in this case.) Once the body is executed for that iteration, the next iteration starts and the variable i then will store the next value from the list (i.e., 20) and the process will repeat until you reach the end of the list.

What if you want to specify certain conditions related to iterations, i.e., on what values your loop should execute? You can do this via for loop guards, as follows, in which you use an if condition in the for loop to specify your conditions.

For instance, to enhance the code example, we can traverse the even elements in the list by using loop guards:

```scala
scala> for (i<-listOfNumbers;if i%2==0){
    println(i)
    }
20
300
-12
```

As you can see in the previous code example:

1. After the `i<-listOfNumbers` we used ; and then an
   `if` condition on the counter variable (i.e., `i` in this
   case) to specify the condition. In this case, we used
   the `%` operator to check the remainder (if the value in
   the variable `i` is divided by 2 and the remainder is 0,
   the number is even).

2. The `for` loop will execute for only those iterations
   where the loop guard condition is true (which in this
   case will be only for even numbers).

You can exploit this capability to write complex logical expressions
using a combination of logical operators like & or | to complete your tasks.

`for` loops are great when you have a list to traverse. When you use
them, you have prior knowledge that the loop will execute/iterate a certain
number of times (N, where N can be the size of the list). However, in many
instances, you may not know beforehand how many times a loop will
execute. Rather, your iteration can be dependent on a condition to be true
with no explicit notion of a collection's traversal. In those scenarios, it
helps to use a `while` loop.

The while Loop

Scala provides another way to use loops, using the `while` statement. In
many scenarios, `for` and `while` are interchangeable.

The overall syntax of the `while` loop is:

```
while(condition){
  body
}
```

Here's a concrete example:

```scala
scala> :paste
// Entering paste mode (ctrl-D to finish)

var i=0
while (i<listOfNumbers.length){
  println(i)
   i+=1
}

// Exiting paste mode, now interpreting.

0
1
2
3
4
i: Int = 5
```

Let's use a step-by-step process to look at what we did in the previous code example:

1. We set our counter variable i to zero prior to starting the loop. Then we used a while loop. In the () of the while loop, we specified a condition which should return a Boolean value (true or false). The loop will run if the condition is true. In this case, it will check the condition if the value of variable i is less than the length of the list.

2. In the body of the function, we specified the logic of while loop i.e. what this loop should do. In this case, it will print each element of the list.

3. We also handled the case so that the loop will terminate at a point. We did this by incrementing the i variable in each iteration. This is important

because if we don't do this, the while loop can run infinitely. This is because the while loop will look at the condition in () and terminate only if this condition becomes false. To make it converge to that case, we need to modify the value of i in each iteration accordingly.

Comparing for and while Loops

You might wonder which of these loop types is best in which scenarios. We address that question in this section. In a while loop, you specify a condition in (), whereas in a for loop, you specify what list to traverse (which can be further qualified with loop guards). Also, in a while loop, you have to handle the iteration logic, i.e., increment your counter variable one by one. Whereas a for loop takes care of this. On this note, a while loop can be used to run indefinitely until a specific condition is true, such as when prompting users for correct passwords again and again until they specify a correct one.

Here's an example of how the password scenario can be implemented in a while loop:

```
scala> :paste
// Entering paste mode (ctrl-D to finish)

var passwd=""
while(passwd != "correctpassword"){
  passwd=scala.io.StdIn.readLine
  println("Enter the correct password")
}

// Exiting paste mode, now interpreting.

Enter the correct password
Enter the correct password
```

```
Enter the correct password
passwd: String = correctpassword
```

In this code example:

1. Upon each iteration of the `while` loop, it will prompt users for their password and will check if it's equal to `correctpassword`. We've used `scala.io.StdIn.readLine` to prompt the user and get input. The value entered by the user will be stored as a `String`.

2. If the value entered does not equal `"correctpassword"`, the condition between `()` will be true and the loop will continue to execute.

3. As soon as the condition within `()` becomes false, it exits.

Here, you haven't specified how many times it will run and thus the `while` loop is better suited for such scenarios. Try implementing the same logic with a `for` loop and see if it's possible.

Breaking a Loop's Iteration

If you want to write logic that allows you to break out of a loop subject to meet some condition, you can use Scala's `scala.util.control.BreakControl`. Here's a simple example of how you can use it:

```
scala> import util.control.Breaks._
scala> :paste
// Entering paste mode (ctrl-D to finish)
var i = 0
while (i<10){
    if (i==7) break
    println(i)
    i+=1
}
```

In this example, you used an `import` statement to import the required modules (`break` in this case) into your program. Then, following the same logic as in previous examples, you specified a variable that served as a counter. In the body of the `while` loop, you specified a condition called (`i==7`), which if true, will execute a `break` statement. The effect of that statement is that it breaks out of the loop, i.e., the loop stops when it encounters a `break` statement.

Thus, when you execute this program, the loop will continue to run until the value of `i` becomes 6. As soon as it becomes 7, the `break` condition becomes true and the loop breaks. The set of statements after the loop's body will continue to execute.

Here's sample output of the code snippet:

```
0
1
2
3
4
5
6
scala.util.control.BreakControl
```

As you can see in the output, Scala REPL also echoes `scala.util.control.BreakControl`, indicating that the loop execution has executed because of the `break` statement.

In this chapter, you learned about yet another commonly used and powerful construct of programming: *loops.* They allow you to repeat execution of your expressions based on certain logic. Practice using them to develop a more thorough understanding of how they work.

EXERCISES

- Load a text file in Scala and see how many times a particular term appears in it. Try implementing this exercise using a for and a while loop.

- Try using loops in a function (e.g., one that prints odd numbers from 0 to 200) and invoke that function in your main program.

- Try assigning a for or while loop to a variable. Understand what happens when you do so.

- Try using two variables in the () after for. Explore the prospects of these and see how you can use them.

- Try nesting one for loop into another.

- Try the good old sorting algorithms (e.g., bubble sort, merge sort) using for and while loops in Scala.

CHAPTER 11

Classes and Packages

We all live in a world composed of different objects. Just pause and look around you. You will probably find objects like chairs, tables, laptops, televisions, and so on. If you put your programming hat on and observe them, you will find that each of these objects has two main parts:

- Each object has some set of properties, such as with a television, it has dimensions (length, width, and height), color, screen type, etc.

- Each object can perform certain functions, such as with a television, you can turn it on, off, increase/decrease volume, change channels, etc.

Objects interact with each other as well. For example, a remote control is an object and it interacts with the television object.

Another aspect inherent in objects is a hierarchy. For example, animals ➤ mammals ➤ humans, dogs, etc.. Animals can be thought as one parent category of living organisms; mammals are a subtype of that, and then humans and dogs are further subtypes of mammals. Many characteristics of mammals are inherited by their subtypes.

Interestingly in the world of programming, under the context of object oriented programming (OOP), we structure our applications using these same principles. In OOP, we identify the objects in our applications using a class and we define their properties, functions, and interactions with them in that class. There is so much in OOP and it is not in the scope of this book

© Irfan Elahi 2019
I. Elahi, *Scala Programming for Big Data Analytics*,
https://doi.org/10.1007/978-1-4842-4810-2_11

to cover OOP in detail. The main purpose of this chapter is to cover some basics of classes and objects, including how they are packaged and how can you use these packages in your programs.

The reason this chapter exists is because when you do Big Data development, specifically in Apache Spark, you use a number of packages to suit your requirements. For instance, if you want to do Apache Spark development in Scala, you use Apache Spark packages and the classes and classes' functions to perform Big Data analytics. Even when you use Scala for general-purpose programming, you still use packages. Chances are that whatever you are trying to implement or build has already been built by someone and is available in Scala. Instead of reinventing the wheel, you can use existing packages. That's why it's important to understand how to use packages.

When you develop executable applications in Scala (and in Big Data analytics using Apache Spark APIs), you structure your programs in such a way that one class becomes the entry point of execution. In those instances, it's always helpful to have a foundational knowledge of classes and objects. Fear not! The notion of object oriented programming may appear to be daunting at first, but as you progress in this chapter, you will find it quite easy—just like many other constructs of Scala that you have studied (and hopefully practiced) in this book.

Classes and Objects in Scala

So, what is a class? So far we have talked about objects only and discussed that objects have properties and that they can perform certain functions. Think of a television as an object, for instance. Let's say that you have a television of a particular make, such as Samsung. This TV model was built by some blueprint or design document. At the Samsung factory, machines and engineers followed those blueprints and design documents religiously to produce the physical TV sets in bulk. You happen to have one of those models in your home.

So we are talking about two things:

- A conceptual specification that details what objects look like. Some examples could be blueprints, design documents, maps, or layouts.

- A concrete and tangible manifestation of that concept, like a house, television, or mobile phone.

In the world of object oriented programming, the first concept in the list is called the *class* and the second one is termed the *object*.

Classes represent at the theoretical/conceptual/logical level what objects contain. Classes don't materialize. They don't exist concretely or physically. Objects are created from classes. (Well, not always, because there is a construct called a singleton object in Scala that doesn't follow this rule, but we'll cover that concept later in chapter later.) Objects possess all the properties and functions defined in classes and they are entities that get initialized and allocated memory when you use them in programming languages.

This concept is illustrated in Figure 11-1.

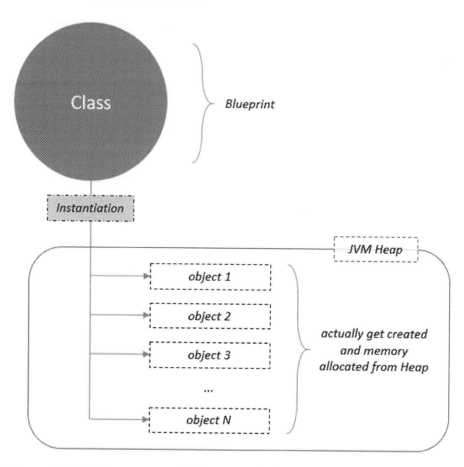

Figure 11-1. *The concept of classes and objects*

Creating Classes and Objects in Scala

As mentioned, this chapter won't go into the details of OOP, but it does share some examples of how to create classes and objects in Scala. In Scala, to create a class, you use the `class` keyword and then specify a class name as follows:

```
class Television {
    //class body
}
```

You use the class keyword, followed by the class name, and then a set of parentheses that contain the class body. As per the naming conventions, class names always start with capital letters. So far, you have created the Television class, but it's empty. It has no properties or functions in it. It, in one sense, is useless. Let's make it useful.

Let's add some details to the Television class. What does a class contain? As mentioned previously, at a very basic level, it contains attributes and functions:

```scala
scala> :paste
// Entering paste mode (ctrl-D to finish)

class Television(val brand:String, val screenSize:Int, val
screenStatus:Boolean) {
    def turnOn = println("Turning on")
    def turnOff = println("Turning off")
}

// Exiting paste mode, now interpreting.

defined class Television
```

Here's what happened in the previous code snippet:

- You defined a Television class.

- The Television class has three attributes (brand, screenSize, and screenStatus, where false means it's off). Think of these as properties or characteristics of that class (and soon to be objects). There can be many other properties of this class, but for the sake of simplicity in this example, we have just three properties.

- The Television class has two functions (turnOn and
  turnOff). Both of these functions perform an action in
  this example; they just print a message onscreen. But
  you can manipulate the screenStatus attribute if you
  want to.

Figure 11-2 highlights the elements that constitute a class.

```
class <className>{

    fields (properties)
    methods (behaviours)


}
```

Figure 11-2. *Elements of a class*

How do you use a class that you've defined? Typically, you will create
objects of it. To define an object in Scala from a class, you use the **new**
keyword:

```
val samsungOled = new Television("Samsung",42,false)
```

Here's what you did in this line of code:

- You created an object called samsungOled from the
  Television class. You used the new keyword for that
  followed by the class name.

- While using the new keyword followed by the class name, you also passed some parameters in the parentheses. Those values are stored against the defined attributes of the class. When an object is created from the Television class, the attributes will be initialized with these values. After this object instantiation, the attribute values will look like this:

  - brand=Samsung

  - screenSize=42

  - screenStatus=false

With that object created, you can now access its attributes. All the attributes and functions that you defined in the original class will be available in the instantiated object.

In Scala, you access elements of an object using the . (dot) operator as follows:

```
scala> samsungOled.brand
res0: String = Samsung

scala> samsungOled.screenSize
res1: Int = 42
```

You can also access the functions that you originally defined in the Television class:

```
scala> samsungOled.turnOn
Turning on

scala> samsungOled.turnOff
Turning off
```

In case you don't remember, you have used the . operator for a while now, for example, when you used the map, filter, and foreach functions on a list, the .length attribute of collections, the .split function of a string, and so on.

In Apache Spark development, you will be using this construct even more. When you use Apache Spark for a lot of cool stuff that is executed in a distributed environment, you will find that your programs will initialize a SparkContext object, just like you initialized a Television object using the new keyword. Then you will use a number of the functions of the SparkContext object to perform a lot of operations, like loading data from a Hadoop distributed filesystem (HDFS), creating accumulator and broadcast variables, and a lot of other cool stuff. Having a strong understanding that objects consist of attributes and functions and that you access them via the . operator is very handy going forward.

Mutating Attribute Values and Caveats

Although it's not too relevant to the scope at hand, I suppose comprehending this topic will help you, especially if you end up creating your own classes. So far, you defined a class and its elements. You created objects of that class and accessed those elements (e.g., attributes and functions).

How about updating the values of attributes of an object? When you create an object, you specify values of its attributes. If you intend to change the values of the attribute for any reason, you can do so, but there are certain caveats. Let's take a look at them:

- When you created the Television class, you defined its attributes with the val keyword. Remember that the val keyword implies immutability. Whenever you create a class with val attributes, you can't update the attribute values. Let's try doing that:

```
scala> samsungOled.screenSize=48
<console>:12: error: reassignment to val
       samsungOled.screenSize=48
                   ^
```

We get reassignment to a val error. You can try this with other attributes. So, what can we do?

When it comes to specifying attributes for Scala classes, you generally have three options:

- Use the val keyword

- Use the var keyword

- Don't use either the val or the var keyword

Each of these options has caveats. Let's learn about each one of them one by one.

Using the val Keyword for Class Attributes

When you use the val keyword, as you've seen before, you can only initialize attribute values at the time of object creation and can't change, mutate, or update their values. In OOP terms, you only get "getters" for attributes and not "setters". Usually in OOP design in other languages, to update or access attribute values, we write some functions in the class, just like the functions we wrote for the Television class (turnOn and turnOff). The function that is used to access attribute values is called getter and the function used to set/update attribute values is called setter (quite intuitive names, aren't they?).

In many instances, you don't need to worry about getters and setters, as Scala creates these for your classes automatically. Specifically, when you specify attributes with the val keyword, you've observe that you don't have the capability to update the attribute values. In other words, you don't get setters for those attributes. You can only access them.

Using the var Keyword for Class Attributes

If you use the var keyword with class attributes, you can both set (or update) and get (or read) attribute values whenever you want. Unlike with the val keyword, even though you initialize attribute values at the start of object creation, you can change them at any time.

From the perspective of getters/setters, you get both of them in such classes. Here's an example of the same class creation but with the var keyword and updating one the attribute values.

```
scala> val lgPlasma=new Television("LG",32,true)
lgPlasma: Television = Television@be694c0

scala> lgPlasma.brand
res13: String = LG

scala> lgPlasma.screenSize
res14: Int = 32

scala> lgPlasma.screenSize=42
lgPlasma.screenSize: Int = 42

scala> lgPlasma.screenSize
res15: Int = 42
```

As you can see, we are not only able to access the attributes but can also change the values.

Using Neither the val Nor the var Keyword for Class Attributes

You also have the option to not use either of these keywords. When you use this option, you can neither access nor update attributes. In other words, you don't getters and setters in such classes.

Try creating the same `Television` class without the `val` or `var` keyword and see if you are able to get/set any of the attributes.

The permutations of using or not using `val` and `var` are highlighted in a structured way in Figure 11-3.

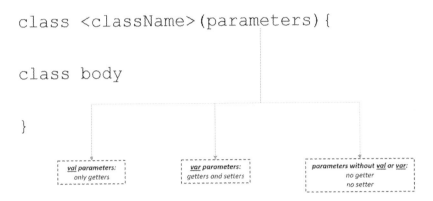

```
class <className>(parameters){


class body


}
```

val parameters:	var parameters:	parameters without val or var:
only getters	getters and setters	no getter
		no setter

Figure 11-3. *Defining class parameters with/without val/var keywords*

Singleton Objects

So far, you have seen that objects are created from classes. You use the new keyword for instantiating an object. Objects are the tangible entities that actually get materialized, whereas classes represent a concept/notion. Now things are going to change a bit, so be ready!

Scala provides another construct, called *singleton objects*. The main proposition of this construct is that these objects aren't created explicitly from classes per se. If you want to use such objects, you don't need to explicitly create a class and instantiate such objects from that class. You get to use such objects directly. Or if put in other way: You don't need to create them using the new keyword.

Though for the geeks, under the hood, Scala does create a class and there is a concept of companion objects at play here, but let's not complicate things.

Singleton objects have all the same characteristics as classes:

- They have attributes and functions

- You access them via the . operator

Let's try creating a singleton object in Scala:

```scala
scala> :paste
// Entering paste mode (ctrl-D to finish)

object DatabaseUtils{
  val databaseName:String = "sample_db"
  val tableName:String = "sample_table"

  def establishConnection = println(s"Establishing connection
to ${databaseName}")
  def closeConnection = println(s"Closing connection to
${databaseName}")
}

// Exiting paste mode, now interpreting.

defined object DatabaseUtils
```

Here's what happened in this code snippet:

- You created an object named DatabaseUtils. Notice that you used the object keyword instead of the class keyword.

- You created attributes of this object (databaseName and tableName).

- You created functions in this object (establishConnection and closeConnection).

This workflow is quite similar to what you did while creating classes in Scala previously.

Furthermore, if you want to use this object, you do as follows:

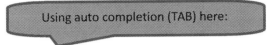

Using auto completion (TAB) here:

```
scala> DatabaseUtils.
closeConnection  databaseName  establishConnection  tableName

scala> DatabaseUtils.databaseName
res16: String = sample_db

scala> DatabaseUtils.establishConnection
Establishing connection to sample_db
```

As you can see in this code snippet:

- The created object has all the attributes and functions that you created. You can retrieve them in Scala REPL by using Tab after typing . after the object name.

- You can use the attributes and function names without explicitly instantiating or creating objects using the new keyword like you did before.

Here are some additional observations:

- Whenever you use this object, it will always have the same attribute values (sample_db for databaseName and sample_table for tableName).

- You can't normally pass attributes as parameters like you did for classes. (Tip: You can achieve that using the apply function to some extent but it's not covered in this book.)

So a natural question would be, when do you use this singleton object in Scala?

- When you want to make your Scala programs executable, you create a singleton object and define the main function with a specific signature. Your application then becomes executable.

- I generally use singleton objects to group utility functions together that don't need child objects. An example is shown in the previous snippet, where we created a utility object to contain database-related functions.

Case Classes

If you talk to Scala developers and ask their opinion of what they love most about Scala, chances are they will mention *case classes*. Case classes are extensively used in Scala and in Apache Spark programming. Let's spend some time looking at them.

Let's work with our current knowledge of classes. When you define a class in Scala, and then create objects out of it, you will find that such classes contain many methods that you didn't specify. For instance, consider Figure 11-4.

```
scala> class Employee(name:String, designation:String, salary:Int)
defined class Employee

scala> val e1 = new Employee("mark","ceo",10000)
e1: Employee = Employee@44e35f

scala> e1.
!=    +    ==          ensuring   equals      getClass    isInstanceOf   notify      synchronized   wait
##    ->   asInstanceOf   eq          formatted   hashCode    ne             notifyAll   toString       ?
```

Figure 11-4. *Default methods in a class*

In Figure 11-4, you created an Employee class, created an object of it, and then tried to access its elements using . and Tab in Scala REPL. You are presented with a series of methods, such as equals, getClass, and toString, to name a few. But you simply created an empty class, right? How on earth did these methods appear?

Referring to the notion that OOP resembles the real world in that it manifests hierarchy in OOP, a class can be a sub-type or a child class of another. Or put other way, the concept of parent and child classes exist in Scala. It's done by virtue of inheritance principles of OOP. By the way, OOP consists of certain principles like inheritance, encapsulation, and polymorphism. As it's not the scope of this book to teach the depths of OOP, so it's suggested you familiarize yourself with these concepts.

Thus, when you create a class, it inherits from the main class java. lang.Object and java.lang.Object actually contains all these functions, which become part of every class that you initialize.

Another OOP concept—you can override many of these functions like toString and equals in accord with your requirements.

This is where case classes shine. Let's explore them.

Case Classes in Practice

So what are case classes? These are classes in Scala that provide boilerplate actions (meaning commonly used code already done for you) for you. Let's look closer at this concept.

When you use case classes, you will find that the behavior of many of their functions (like toString and equals) is quite different from the simple classes in Scala. They are overridden for you. Their behavior is modified to assist in their usage in Big Data analytics. This point will become clear in a while.

When you use a case class, you don't need to initialize the objects with the new keyword. This is similar to how singleton objects work.

Consider the following example:

```scala
scala> case class EmployeeCaseClass(val designation:String,
val company:String)
```

That's how you define a case class. You use "case class" keyword and then class name and then define the parameters. If required, you can define your methods within the body of the case class as well.

Also, when it comes to creating individual objects out of the case class, you don't initialize them with the new keyword. Remember the *companion classes* term I used when describing singleton objects? It's because of that. At this stage, it's enough that you know this concept.

In Apache Spark, case classes are used extensively. As an example, there are two types of (main) data structures in Apache Spark: RDD and dataframe. RDD are like Scala collections (with a number of characteristics like immutable and distributed) and dataframes are like tables or spreadsheets. You extensively use case classes when you want to convert data from RDD to dataframe (RDD and dataframe are two of the data structures available when you use Apache Spark for distributed computing. We'll cover these concepts in a later chapter.) Trust me that this operation is quite extensively used in Apache Spark. Having a hands-on understanding of the case class is in line with your goal of using Scala for Big Data analytics.

Referring to the aforementioned notion that many of the behaviors of case class methods are modified, let's dig into this concept further.

Equality Checks in Classes

Let's say you want to compare two classes (or more specifically the objects of those classes) to determine whether they are equal. One way to do this is to check individual fields of the objects against each other. That's what you normally do if you create simple classes (ones that are not case classes).

This is the norm in other languages like Java. For instance, consider the following example:

```
scala> class Employee(val designation:String, val
company:String)
defined class Employee

scala> val e1=new Employee("engineer","facebook")
e1: Employee = Employee@a5cb99

scala> val e2=new Employee("engineer","facebook")
e2: Employee = Employee@6d58cd
```

In this code snippet:

- We created a class called Employee.

- We created two objects of this class (e1 and e2).

The attribute values of these two objects are the same—the designation in e1 and e2 is "engineer" and so is company ("facebook"). But if you compare the equality of these two objects:

```
scala> e1.equals(e2)
res0: Boolean = false
```

You instantly note that you get false in return. You know that from the attribute values' point of view, both objects are the same. But when you use the equals method, instead of checking attribute values, it checks the memory address of the object reference (specifically the starting address location in the heap memory). As each object gets a different memory location, it's returned as false.

As in analytics, we do a lot of computations and many of those relate to equality checks. So if you are performing analytics at scale and create your own type (via case classes) and you want to compare them, that's where case classes come in handy.

Consider the same example, but implemented using case classes as follows:

```
scala> case class EmployeeCaseClass(val designation:String,
val company:String)
defined class EmployeeCaseClass

scala> val ec1=EmployeeCaseClass("engineer","facebook")
ec1: EmployeeCaseClass = EmployeeCaseClass(engineer,facebook)

scala> val ec2=EmployeeCaseClass("engineer","facebook")
ec2: EmployeeCaseClass = EmployeeCaseClass(engineer,facebook)
```

In this code snippet:

- We created a case class (EmployeeCaseClass) and defined its parameters.

- We then created two objects of this class (ec1 and ec2). Note again that we didn't use the new keyword here, as we do with simple classes.

Now let's try to check the equality of these objects to verify whether case classes live up to the expectation or not:

```
scala> ec1.equals(ec2)
res1: Boolean = true
```

As expected, it says true. Thus, when you compare case classes (or their objects), it doesn't compare their memory addresses. Rather, it compares individual attributes, which is pretty handy in many circumstances.

Another difference between a case class and a simple class is when you invoke the toString method on both:

```
scala> e1.toString
res2: String = Employee@a5cb99
```

```
scala> ec1.toString
res3: String = EmployeeCaseClass(engineer,facebook)
```

As you can gather from this code, if you invoke the toString method on a simple class, it doesn't return anything intuitive. Whereas in a case class, it returns something from which you can determine which attributes values were used for its initialization.

Case Classes and Collections Together

Before we move to other aspects of classes, I wanted to highlight how you can use case classes with collections.

In Chapter 9, you created collections of different native types, like collections of Integer, String, etc. How about a collection of a case class?

For example, if you want to create a collection that holds data about the employee class, how would you do that?

First let's create a collection that holds such data:

```
scala> val employees = List("engineer,facebook","manager,
facebook","associate,facebook")
employeeData: List[String] = List(engineer,facebook,
manager,facebook, associate,facebook)
```

Let's also create a case class to hold the employee data:

```
scala> case class EmployeeData(designation:String,
company:String)
defined class EmployeeData
```

Now let's put our knowledge of collections into practice and iterate through the list of strings to create a list of our case class:

```
scala> val employeeList =  employees.map(x=>x.split(",")).map(x
=>EmployeeData(x(0),x(1)))
```

```
employeeList: List[EmployeeData] = List(EmployeeData(engineer,
facebook), EmployeeData(manager,facebook), EmployeeData
(associate,facebook))
```

employeeList is a list of objects in which each object is a case class. This can become quite handy. For instance, if you index the first element of this list, you get an object of a case class:

```
scala> employeeList(0)
res0: EmployeeData = EmployeeData(engineer,facebook)
```

You can use that to further access the individual attributes:

```
scala> employeeList(0).designation
res1: String = engineer
```

If you want to get (or print) designations of your employees, you can use something like this:

```
scala> employeeList.map(x=>x.designation)
res2: List[String] = List(engineer, manager, associate)
```

As you can gather, when you combine case classes with lists, you can do a lot of powerful data manipulation. You will find that in Apache Spark programming, this notion is used quite a lot.

Classes and Packages

So far we have learn about classes, including learning how to create them and use them via objects. As mentioned, OOP principles are related to real-world objects and in real-world objects, there is almost always the concept of hierarchy. A similar notion exists in OOP—one class can be a sub-class of another. Programmers use this capability to optimally structure their programs, which also fosters reusability.

However, there is another aspect of hierarchy in classes. When you use or create classes, you'll find that classes are grouped into something called a *package*. Packages are special entities in Scala used to contain classes. Generally, package names have an intrinsic hierarchy in them that relates to the organizational domain name. For example, if you use the Apache Spark package, it's structured as `org.apache.spark`; if you use the Microsoft SQL Server JDBC driver, you use the `com.microsoft` package; if you want to use Cloudera's Impala driver, you use the `com.cloudera` package. Package names are usually the reverse of a company's domain name (for example, package name: `com.microsoft` and domain name: `microsoft.com`).

Avoiding Name-Space Collisions

Packages, among many other benefits, help developers avoid name-space collisions. For example, when you use Scala REPL, you can access the `Array` collection there (specifically `scala.Array`). Now, if you do JDBC programming—writing programs to connect and interact with databases—you will find that you will use the `java.sql` package quite a lot. In that package, there is an `Array` class as well, called `java.sql.Array`. As both `Array` classes are part of different packages—one belongs to `scala.Array` and the other to `java.sql.Array`—and if you import them properly in your program, you can avoid name-space collision. That means when you refer to `Array` in your programs, Scala will know whether you are referring to `scala.Array` or `java.sql.Array`.

If you reflect, you will find that you have already been using packages and their respective classes in this book. Remember when you used mutable and immutable collections specifically in the form of mutable and immutable maps? You used a specific package for both of them.

You don't initialize packages per se as you define classes or singleton objects. Rather, as you'll find in the upcoming chapters when you write classes, you define, mostly at the start of programs, to which package this class belongs. By doing so, the class becomes part of that package and you can reuse that class in other parts of the program.

Importing Packages

When you write Scala programs, you have to rely on many packages that either you created or that were developed by others. By default, when you write programs, those packages aren't loaded automatically in your namespace. If you want to access a particular package or a particular class in a package, you won't be able to do so unless you do something about it. For example, if you want to read a file in your local filesystem in your program, Scala provides the `fromFile` function for this purpose and this function is part of the `scala.io.Source` class.

If you want to use `Source.fromFile` in your program right away, you won't be able to do that because this class isn't loaded into your JVM environment (or into Scala REPL if you are using that). Technically speaking, this class isn't on the default classpath of JVM. What's a classpath, you may ask? It's Java-specific jargon that refers to the path in your local system where Java looks for classes when you try to use them.

If you try to use the `Source.fromFile` function in your Scala REPL without importing it properly, here's what'll happen:

```
scala> val fileData=Source.fromFile("C:\\path_to\\sample_file.txt")
<console>:14: error: not found: value Source
       val fileData=Source.fromFile("C:\\path_to\\sample_file.txt")
```

You will get an error in which Scala is complaining that it can't find Source, meaning this particular object, although it exists, isn't accessible in your environment.

How can you go about addressing errors like these? You have two choices:

- Specify the full name of the class, including the package to which it belongs, that you want to use whenever you need:

```
scala> val fileData = scala.io.Source.fromFile
("valid path to a file")
fileData: scala.io.BufferedSource = non-empty iterator
```

   If you do this, you will have to type the full name again and again, which can lead to typos and will needlessly increase the verbosity of your code.

- You can alternatively import that class in your environment and use it right away, without the need to fully specify the name:

```
scala> import scala.io.Source
import scala.io.Source

scala> val fileData=Source.fromFile("valid path to a file")
fileData: scala.io.BufferedSource = non-empty iterator
```

A couple of observations here:

- You imported the required class (scala.io.Source) using the import keyword.

- Once you import it, you use the class without using its full name (Source.fromFile() instead of scala.io.Source.fromFile()).

You can even do the following:

```
scala> import scala.io.Source._
import scala.io.Source._

scala> val fileData = fromFile("valid path to a file")
fileData: scala.io.BufferedSource = non-empty iterator
```

In this attempt, the following things happened:

- You used the import command and imported scala.io.Source._. What does import scala.io.Source._ mean? It means to import everything that's in the scala.io.Source class. Since fromFile was part of this class, you can use this function directly in your program and you didn't even have to specify Source.fromFile like you did before.

- If you used the Java language, then import scala.io.Source._ is equivalent to import scala.io.Source.* in Java.

There are a couple of caveats related to import statements that deserve an explanation:

- As you've seen before, if you use the ._ notation, you import all entities in a class or package. Those entities can be a class in a package or a function (or fields) in a class. It's generally considered bad practice to use this notation because it results in importing all the entities (e.g., classes, functions, etc.) in your program, even the ones you might not use. By doing so, this can result in a name-space collision.

- You can import an individual class using the import statement. Like you did before when you used import scala.io.Source.

- You can import multiple individual classes/objects in a package using the import statement. To do this, you use brackets {} with the import statement:

```
scala> import scala.io.{Source,StdIn}
import scala.io.{Source, StdIn}
```

- While importing, you can assign a label for the class for your use. This can become quite handy to avoid name-space collision. As mentioned, when you use Scala REPL, the Array class is loaded automatically. If you want to use the java.sql.Array class and import using this:

```
import java.sql.Array
```

And if you use Array, it will tend to use java.sql. Array instead of scala.Array (which Scala imports for you by default in Scala REPL). Here's an example:

```
scala> import java.sql.Array
import java.sql.Array

scala> val numberArray = Array(1,10,-100)
<console>:18: error: class java.sql.Array is not a value
       val numberArray = Array(1,10,-100)
                         ^
```

To avoid such scenarios, you can import java.sql.Array and assign it a different label. As a result, you can continue to use scala.Array and java.sql.Array simultaneously in your program without any conflicts.

```
scala> import java.sql.{Array=>SqlArray}
import java.sql.{Array=>SqlArray}
scala> val numberArray = Array(1,10,-100)
numberArray: Array[Int] = Array(1, 10, -100)
```

In this code example, when you used Array, the Scala interpreter will know right away what type of array are you referring to because you specifically created an alias of java.sql.Array as SqlArray. Thus, when you used Array in your code, it didn't conflict with java.sql.Array and it created an array of the right type (scala.Array).

This concludes this chapter. It's important to understand how to import your required classes from packages in your program so that you avoid name-space collisions, along with improving verbosity and readability of your code.

EXERCISES

- Try importing Java libraries in Scala—e.g., Java's date libraries—and try using them in your program.

- If you are using Linux, try executing the shell script from your Scala program. You will have to import a specific package to do so.

- Research the maven repository. What is it, what does it contain, and how do Scala and Java programmers use it?

- Try creating two classes—Television and PowerSupply—
  to model a television set with a power supply component. Try
  to create a Television class where one of its attributes is
  PowerSupply. Can you do that? On the same note, search for
  "composition in object oriented programming".

- Research the dependencies in the context of libraries. For
  example, if your program intends to use a specific library that
  doesn't come out of the box, what can you do about it? Try using
  the spray-json library, which is great for parsing JSON files
  and converting to case classes in Scala. You can't import it as it
  is. Try to sort out how you can import it.

CHAPTER 12

Exception Handling

It's a cliché that life is unpredictable. At any point in life, the unexpected can happen and in order to survive the surprises of life, it's important to be well prepared.

The same philosophy, to some extent, holds true with computer programs. At runtime or compile time, unexpected issues can happen. They can be due to a number of reasons, including:

- Your program was expecting a specific input, whereas the input provided was a different format.

- The program depended on an external system, such as a database to be available, and that system was down or unstable at the time program wanted to interact with it.

- Your program was trying to access an object that wasn't defined before (such as accessing an index of a collection that's more than the length of the collection).

These scenarios can vary. When such issues appear, you have two options:

- You don't handle them. You let the programs depend on "ideal conditions" and assume that everything will be favorable.

- You handle exceptions as and when they appear and take actions accordingly.

© Irfan Elahi 2019
I. Elahi, *Scala Programming for Big Data Analytics*,
https://doi.org/10.1007/978-1-4842-4810-2_12

Good programmers follow the second approach. If you are getting started in Scala, you should think with this mindset as well. Specifically, as you will be using Big Data technologies that rely on distributed systems and processes and on external systems a lot, the probability of exceptions increases to some extent. Thus, if you are using Scala, it makes sense to be well versed in exception handling so that your programs won't crash and won't land you in trouble.

Fundamentals of Exception Handling in Scala

In Scala, if you suspect that a particular expression will result in an exception, you must use and write certain Scala programming language constructs to handle those exceptions.

Figure 12-1 shows the overall mechanics of exception handling in programming in general.

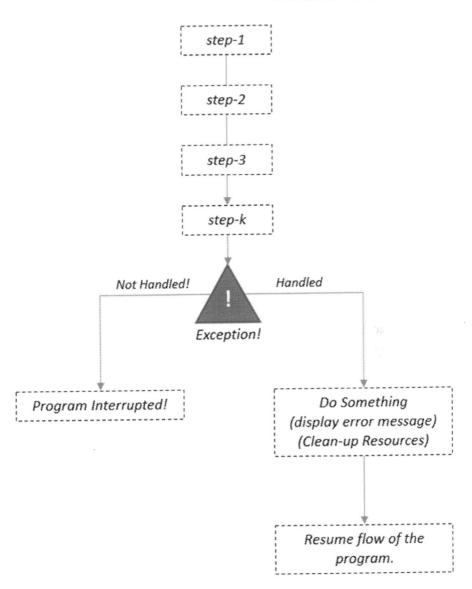

Figure 12-1. *General concept of exception handling in programming languages and in Scala*

The most straightforward way to handle exceptions in Scala is to use try-catch blocks. The format looks like this:

```
try {
    //the expressions that can cause exceptions
} catch {
    //your logic to handle exceptions
}
```

The try-catch block construct isn't exclusive to Scala. It's available in other languages as well (with some variations of course).

However, Scala, being a super-set of Java and an amazing language by itself, provides more powerful constructs to handle exceptions optimally. Remember pattern matching? Recall the constructs that you studied in Chapter 6 that allowed you to replicate the behavior of switch statements as used in other languages, and allowed you to achieve multiple nested conditions in a clean and powerful way. You're going to use similar concepts here for exception handling.

Let's consider a basic example:

```
scala> val aList = (1 to 20).toList
aList: List[Int] = List(1, 2, 3, 4, 5, 6, 7, 8, 9, 10, 11, 12, 13, 14, 15, 16, 17, 18, 19, 20)
```

We defined a list consisting of 20 elements. What would happen if you try to access the 22nd element of this list? Exception!!

```
scala> aList(22)
java.lang.IndexOutOfBoundsException: 22
  at scala.collection.LinearSeqOptimized.
apply(LinearSeqOptimized.scala:63)
  at scala.collection.LinearSeqOptimized.
apply$(LinearSeqOptimized.scala:61)
  at scala.collection.immutable.List.apply(List.scala:86)
  ... 28 elided
```

You get a specific type of exception here, known as IndexOutOfBounds Exception. The name itself is quite intuitive, as you were trying to access an index of a collection that was out of bounds (i.e., more than the length of the collection). When you run this program, your program's execution will be interrupted and you will get this ugly stack trace on stdout, which isn't a decent way to inform us of the issue, indeed. There has to be a better way.

Our expression aList(22) is susceptible to an exception, right? Such culprit statements should be wrapped with a try block. The resulting probable exceptions are caught in an accompanying catch block, where you use a series of case statements to specify how you want the exceptions to be handled. Something like this:

```scala
scala> :paste
// Entering paste mode (ctrl-D to finish)

try{
  aList(22)
} catch {
  case _: Throwable => "exception"
}

// Exiting paste mode, now interpreting.

res10: Any = exception
```

Here's what the previous code snippet is doing:

- You used the paste mode of REPL so that you could execute a block of expressions at once.

- You used the try keyword and then defined a code block with brackets {}.

- Within the try code block, you specified the expression that you wanted to execute and indicated the usual suspects for generating exceptions.

- You then wrote a `catch` block. `try` is always accompanied by a `catch` block. Remember that.

- In the `catch` block, you can specify a series of `case` statements in a similar way you used them in pattern matching scenarios. In this example, we just used one `case` statement.

- In our `case` statement, we used `_`. Remember what it does? After `case`, you can specify a temporary variable that you then can use in the `case` block. Here you used `_`, which means you don't want to define a variable.

- You defined the type that you want to match via `case _:Throwable`. Throwable is a JVM-specific construct (or more specifically it's a class) that represents general exceptions. There are also concrete classes of this class. In your `case` statement, you are matching for a general occurrence of an exception. You aren't looking for a particular type of exception. It's like a catch-all scenario; it matches if any type of exception happens.

- If the exception happens, that `case` statement will be matched and whatever is to the right of `=>` will be executed. This is where your logic for handling the exception will go. In our example, we just returned the `"exception"` string, which is why, when you execute this program, this is returned (well, type-cast to its parent type `Any`).

So this is the overall anatomy of exception handling in Scala in its very basic form. Let's make this a bit better and more involved, shall we?

```
scala> :paste
// Entering paste mode (ctrl-D to finish)

try{
  aList(22)
} catch {
  case x: IndexOutOfBoundsException => "index out of bounds
exception"
  case _: Throwable => "exception"
}

// Exiting paste mode, now interpreting.

res12: Any = index out of bounds exception
```

Most of the stuff in this example will be clear to you, but I just want to bring your attention to a few specific points:

- As I mentioned, you can write a series of case statements. That's what you did here.

- In the first case statement, you looked for a specific type of exception, i.e., IndexOutOfBoundsException. In the next case statement, you looked for a more general exception condition. This is generally how programmers write handlers for exception handling. They start with a specific type and, as they write more case statements, they make their handlers more and more generic.

- When you executed this snippet of code, the first case statement was executed. This is the one that was matched.

- In the first case statement, you used the x variable. By doing so, we have an additional capability to use the x variable in the body of the case statement (in this example it's not used that way, but we'll use it shortly in this chapter).

Here's another twist to exception handling:

```
scala> :paste
// Entering paste mode (ctrl-D to finish)

try{
  aList(22)
} catch {
  case x: IndexOutOfBoundsException => throw new
IndexOutOfBoundsException
  case _: Throwable => "exception"
}

// Exiting paste mode, now interpreting.

java.lang.IndexOutOfBoundsException
  at .liftedTree1$1(<pastie>:16)
  ... 36 elided
```

In this example, in the first case statement block after =>, you threw the exception yourself. For example, if an error is critical and you don't want the program to continue working after it has been encountered, you throw exceptions. One scenario could be with a data processing pipeline, where you are reading from source files and writing to databases. Say you are doing data quality checks on a file and find issues in the data (e.g., data type mismatches), one approach is to throw an exception so that you don't insert dirty records into the final database.

If you need to explicitly throw exception yourself, you can use this form:

```
throw new <exception type>
```

Let's look at another example:

```
scala> :paste
// Entering paste mode (ctrl-D to finish)

try{
  aList(22)
} catch {
  case x: IndexOutOfBoundsException =>
    println("Printing error stack trace for better trouble-
    shooting")
    x.printStackTrace()
  case _: Throwable => "exception"
}

// Exiting paste mode, now interpreting.

Printing error stack trace for better trouble-shooting
java.lang.IndexOutOfBoundsException: 22
        at scala.collection.LinearSeqOptimized.apply
        (LinearSeqOptimized.scala:63)
        at scala.collection.LinearSeqOptimized.apply$
        (LinearSeqOptimized.scala:61)
        at scala.collection.immutable.List.apply(List.scala:86)
        at $line26.$read$$iw$$iw$.liftedTree1$1(<pastie>:14)
        at $line26.$read$$iw$$iw$.<init>(<pastie>:13)
        at $line26.$read$$iw$$iw$.<clinit>(<pastie>)
        at $line26.$eval$.$print$lzycompute(<pastie>:7)
        at $line26.$eval$.$print(<pastie>:6)
        at $line26.$eval.$print(<pastie>)
```

```
at sun.reflect.NativeMethodAccessorImpl.invoke0
(Native Method)
at sun.reflect.NativeMethodAccessorImpl.invoke
(Unknown Source)
at sun.reflect.DelegatingMethodAccessorImpl.invoke
(Unknown Source)
at java.lang.reflect.Method.invoke(Unknown Source)
at scala.tools.nsc.interpreter.IMain$ReadEvalPrint.
call(IMain.scala:735)
at scala.tools.nsc.interpreter.IMain$Request.loadAndRun
(IMain.scala:999)
at scala.tools.nsc.interpreter.IMain.$anonfun$interpret$1
(IMain.scala:567)
at scala.reflect.internal.util.ScalaClassLoader.asContext
(ScalaClassLoader.scala:34)
at scala.reflect.internal.util.ScalaClassLoader.asContext$
(ScalaClassLoader.scala:30)
at scala.reflect.internal.util.AbstractFileClassLoader.
asContext(AbstractFileClassLoader.scala:33)
at scala.tools.nsc.interpreter.IMain.loadAndRunReq$1
(IMain.scala:566)
at scala.tools.nsc.interpreter.IMain.interpret
(IMain.scala:593)
at scala.tools.nsc.interpreter.IMain.interpret
(IMain.scala:563)
at scala.tools.nsc.interpreter.ILoop.$anonfun$paste
Command$11(ILoop.scala:816)
at scala.tools.nsc.interpreter.IMain.withLabel
(IMain.scala:112)
at scala.tools.nsc.interpreter.ILoop.interpretCode$1
(ILoop.scala:816)
```

```
    at scala.tools.nsc.interpreter.ILoop.pasteCommand
    (ILoop.scala:822)
    at scala.tools.nsc.interpreter.ILoop.$anonfun$standard
    Commands$10(ILoop.scala:190)
    at scala.tools.nsc.interpreter.LoopCommands$Line
    Cmd.apply(LoopCommands.scala:154)
    at scala.tools.nsc.interpreter.LoopCommands.colonCommand
    (LoopCommands.scala:114)
    at scala.tools.nsc.interpreter.LoopCommands.colonCommand$
    (LoopCommands.scala:112)
    at scala.tools.nsc.interpreter.ILoop.colonCommand
    (ILoop.scala:43)
    at scala.tools.nsc.interpreter.ILoop.command
    (ILoop.scala:752)
    at scala.tools.nsc.interpreter.ILoop.processLine
    (ILoop.scala:456)
    at scala.tools.nsc.interpreter.ILoop.loop(ILoop.scala:477)
    at scala.tools.nsc.interpreter.ILoop.process
    (ILoop.scala:1069)
    at scala.tools.nsc.MainGenericRunner.runTarget$1
    (MainGenericRunner.scala:82)
    at scala.tools.nsc.MainGenericRunner.run$1
    (MainGenericRunner.scala:85)
    at scala.tools.nsc.MainGenericRunner.process
    (MainGenericRunner.scala:96)
    at scala.tools.nsc.MainGenericRunner$.main
    (MainGenericRunner.scala:101)
    at scala.tools.nsc.MainGenericRunner.main
    (MainGenericRunner.scala)
res16: Any = ()
```

Wow, quite a lengthy error stack trace, isn't it?

Here's what we did differently in the previous code snippet:

- In the first case statement, you used multiple statements in that case block. This is not unusual. There is generally more than one statement in the block and you don't necessarily need to surround them in brackets {} specifically in such instances, like when using case statements.

- In the first case statement, we matched for a specific exception type, called IndexOutOfBoundsException. The second case statement match is done for a general type of exception.

- We deliberately executed an expression (aList(22)) against which we got the IndexOutOfBoundsException exception. As a result, the first case statement was matched. Thus, its code block was executed and the second case statement didn't execute.

This is highlighted further in Figure 12-2.

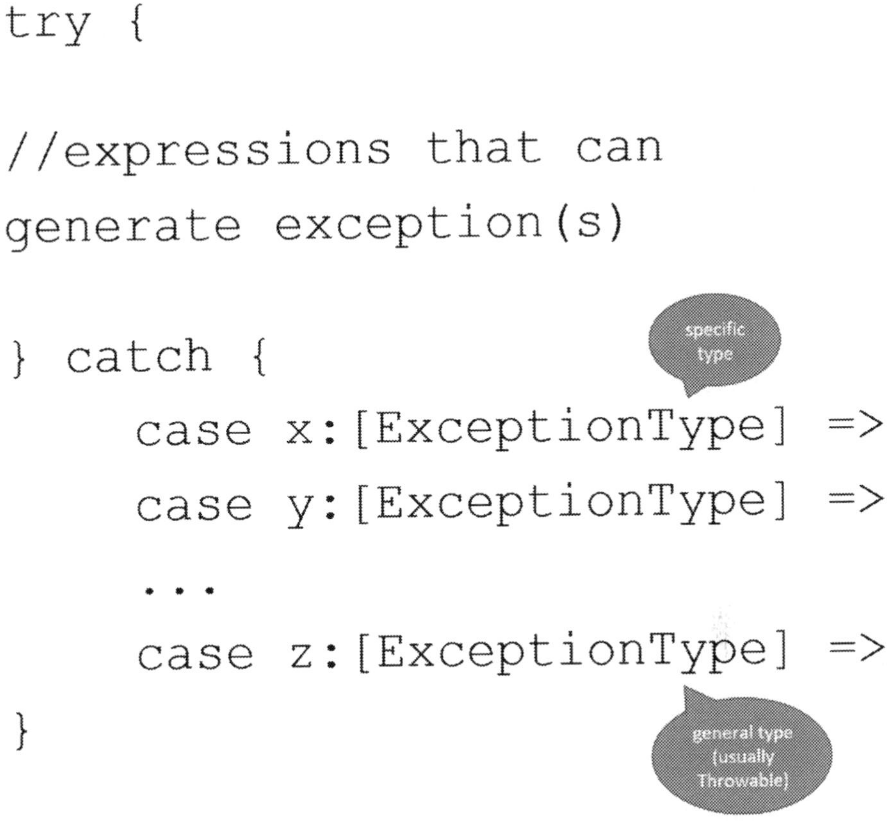

```
try {

//expressions that can
generate exception(s)

} catch {
    case x:[ExceptionType] =>
    case y:[ExceptionType] =>
    ...
    case z:[ExceptionType] =>
}
```

specific type

general type (usually Throwable)

Figure 12-2. *Using try-catch blocks to handle exceptions in Scala*

Implications of Type Inference and Exception Handling

As a refresher, you use code blocks when the intent is to assign a value to a variable after the statements in the code block have been executed. Similarly, in many scenarios, you will want to execute expressions and ensure that you use exception-handling constructs if anything bad happens with those expressions. The result of that expression should be assigned to a variable. It's a normal expectation; however, when you do that in Scala, there are certain implications that I want to highlight.

Consider the following snippet of code:

```scala
scala> val inputIndex = scala.io.StdIn.readLine().toInt
```

```scala
scala> :paste
// Entering paste mode (ctrl-D to finish)
val theElement = try{
  aList(inputIndex)
} catch {
  case x: IndexOutOfBoundsException =>
    println("Printing error stack trace for better trouble-shooting")
    x.printStackTrace()
  case _: Throwable => "exception"
}
```

```scala
// Exiting paste mode, now interpreting.
```

```scala
theElement: Any = 20
```

So here's what we did in the previous code snippet:

- When the program runs, it will prompt users for input. If, for instance, the user inputs 19, that value will be stored in the inputIndex variable, which is then used in indexing aList. If the indexInput value is 19, it will access the 20th element of the list (which has 20 elements). This operation will not result in an exception. But, being a thoughtful programmer, you wrapped this operation in an exception-handling block by using try-catch.

- In the case statements, you used the same set of conditions and code blocks as before.

- What you did differently is assigned the value of the whole exception block to a variable called theElement.

- Your expectation might be that if you do a normal operation like accessing the 20th element of the list using aList(19), it will not result in an exception and the value should be stored with its right data type (Integer). In this scenario, Scala did return the value as 20, but it type-cast it to Any instead of Integer. This type-casting can have implications down the road. Perhaps you developed your program to perform numerical operations on this variable, but now you have variable of type Any here. The reason that it's doing this is that in one of your case statements, you are returning a string ("exception") and thus Scala reverted to the higher data type.

Even if you try to specify the type of this theElement variable to be an Integer as follows, it won't help:

```scala
scala> val inputIndex = scala.io.StdIn.readLine().toInt
scala> :paste
// Entering paste mode (ctrl-D to finish)

val theElement:Int = try{
  aList(inputIndex)
} catch {
  case x: IndexOutOfBoundsException =>
    println("Printing error stack trace for better trouble-shooting")
    x.printStackTrace()
  case _: Throwable => "exception"
}

// Exiting paste mode, now interpreting.

<pastie>:17: error: type mismatch;
 found    : Unit
```

217

```
 required: Int
x.printStackTrace()
                  ^

<pastie>:18: error: type mismatch;
 found    : String("exception")
 required: Int
case _: Throwable => "exception"
                     ^
```

Instead, it will result in an error for reasons related to type-mismatch.

Using Try, Catch, and Finally

You will often be working with external resources in your program, for instance when reading from or writing to a file/database/message queues. When you interact with those resources, you establish a connection to them and then write data via that established connection. As a best practice, it is recommended that, when you are done interacting with those external resources, you close the connection (or any other objects that you may have initialized). There is a cost associated with each connection (e.g., when you establish a connection to a database, it reserves a connection slot) and if you continue to establish connections without closing, it can have performance impacts on the external systems. This notion becomes even more important when you are writing expressions that throw exceptions. In your program, you may have written the logic to close a connection, but if your program gets interrupted because of exceptions, that logic won't be executed and the resources will not be released.

To handle this problem, programming languages, including Scala, provide constructs like try, catch, and finally. The main premise of try-catch remains the same, but the addition of the finally block ensures that the code in it is always executed, even when exceptions occur. This ensures that you don't run into the resources exhaustion issue we discussed previously. In the finally block, you generally write expressions related to releasing resources, such as closing connections.

Here is an example that elaborates the point further:

```
import java.io.{File, BufferedWriter, FileWriter}
val fileContent = scala.io.StdIn.readLine()
val textFile = new File("valid path to a file")

try {
    val buffWriter = new BufferedWriter(new FileWriter(textFile))
} catch {
    case x:java.io.IOException => println("Issues in writing
file. Check if you have permissions to write to the location or
if a directory with the same name exists")
                                throw new java.io.IOException

}
try {
    buffWriter.write(fileContent)
} catch {
    case x:java.io.IOException => println("Writing to the file
failed. Check if you have the permissions to write to file")
                                throw new java.io.IOException
} finally {
    buffWriter.close()
}
```

In this code example:

- We used Java libraries (File, BufferedWriter, and FileWriter) that allow you to write to a file.

- First, we got input from a user about the content that needs to be written. Then we initialized an object of the java.io.File class and specified the path where the file will be created and where the data will be written.

- Then we used `try-catch` blocks to attempt to write to a file. In such instances, it helps to resort to online documentation to assess which exceptions a particular class or method can throw. For instance, Figure 12-3 shows the details of the `java.io.FileWriter` class. If you observe the `throws` section, it highlights the type of exception that this class can throw when initialized.

Figure 12-3. *Documentation of the java.io.FileWriter class with emphasis on the type of exceptions it can throw*

- As you can see, it can throw the `java.io.IOException` exception, so the code is written to catch those expressions.

- At the end of code, the `finally` block is used to close the `BufferedWriter` object. That is like a connection stream to the file where data is written.

The key takeaway here is that you should use exception handling in your code. However, when you use exception handling, you should be vigilant of the different caveats of using exception handlings in Scala, as discussed in this chapter.

EXERCISES

- When you use a function from a module, try to observe what type of exception the function can throw in the Scala documentation online. Try to write programs in such a way that you handle all of those exceptions.

- Explore the benefits of using `scala.util.` `{Try,Success,Failure}` for exception handling.

CHAPTER 13

Building and Packaging

Doesn't it feel great that you are nearing your goal to learn Scala for Big Data analytics? With such an ambitious goal, you need to orient yourself according to the development patterns. You need to follow the development lifecycle currently practiced in the development community. You need to be exposed to another dimension in the development endeavors, which will enable you to extend your development efforts beyond Spark Shell. In a nutshell, now is the right time to pivot to topics that relate to building and packaging your Scala code.

Let's put things in perspective. So far, you've been using Scala REPL for all of your development in this book. Scala REPL is an amazing tool and it really fosters productivity of a programmer and brings you up to speed. You get your hands dirty with the quirks and constructs of programming. You can use it to quickly test, learn, and prototype something.

However, in production environments, your logic/program/ expressions won't be executed in Scala REPL. It won't be like that if you want to do something with your Scala code, such as count the frequency of a word in some documents. You won't invoke the Scala shell and pass your commands one by one to it. Also, the distribution of code is an equally important aspect that a good developer needs to consider. Generally there are multiple environments in professional settings—development, test, and production. Generally speaking, developers operate in a development environment and then their code is tested in test environments. If the tests pass, the code is deployed into a production environment. If your code is

© Irfan Elahi 2019
I. Elahi, *Scala Programming for Big Data Analytics*,
https://doi.org/10.1007/978-1-4842-4810-2_13

packaged in such a way that it is easily distributed, executed, and tested in different environments, it refines the overall development pipeline and aligns with the DevOps practices prevalent in today's industry.

The Scala Development Lifecycle

The approach that's generally followed in this lifecycle is as follows:

1. You use Scala REPL in the aforementioned capacity, for rapid learning or prototyping.

2. You then create Scala applications that are properly structured in the form of classes and packages.

3. Your code is generally divided into a number of Scala files that constitute your classes and packages as a whole.

4. You define and use any additional modules you need in your Scala applications (e.g., a library for parsing JSON, a library to enable you to interact with Microsoft SQL Server via JDBC, or a library that allows you to use Apache Spark API). This is called dependency management.

5. Once you structure your app and manage your dependencies, you perform the following steps:

   - Compile

   - Build

   - Test

   - Package

   - Deploy

As a result, your Scala application becomes compiled, tested, and packaged in the form of a JAR (Java Archive). JAR is a file format mostly used for packaging your code along with its dependencies (e.g., libraries), metadata, and any other required resources. Java and Scala programs are packaged in JAR form for distribution and deployment.

These steps may appear to be nebulous at the start and that's the goal of this chapter—to explain these concepts and strengthen your skillset in these areas.

The Scala Development Lifecycle in Action

Let's start with a goal—you want to create an executable Scala application packaged in the form of JAR. By executable, I mean that the JAR is runnable: it does something. Let's start with a very basic example: a Hello World example. You've already created one in Scala REPL, but let's do it in the way professional developers do.

This will lay the foundations of Apache Spark development as well. To achieve this goal, you need the following:

- An IDE. We'll use IntelliJ, but we'll start with a text editor and you can use any text editor of your choice, even one as simple as Notepad. Better ones include Sublime Text Editor, Visual Studio Code, and Notepad++. The choice is yours. I won't go into the details of setting up or installing a text editor, as it's pretty trivial.

- A build tool. In Scala, the de facto tool is Scala Build Tool (SBT), which is genuinely amazing. This tool allows you to do a lot of things, not limited to:

  - Manage dependencies

  - Compile your code

- Run unit tests

- Package your code in the form of JAR

We'll see each of these actions in a bit. We won't cover unit tests in this book as they demand substantial discussion on their own.

Scala Build Tool (SBT)

The consensus is that you'll need SBT. We've been using Windows in this book to date, so let's stick to that.

To install SBT, navigate to the following website and download the Windows MSI Installer:

```
https://www.scala-sbt.org/1.x/docs/Installing-sbt-on-Windows.html
```

This installer is shown in Figure 13-1.

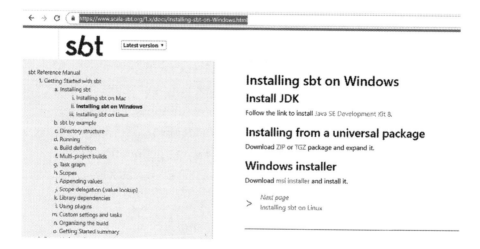

Figure 13-1. *Installing Scala Build Tool on Windows*

Once the installer is downloaded, the rest of the installation is pretty straightforward. It will ask for an end user license agreement, components, and location to install. At this stage, you can proceed with the defaults. Figure 13-2 shows how the installation wizard looks.

Figure 13-2. *Installation wizard of SBT on Windows*

Once the installation is complete, you should be ready to use SBT.

Using SBT on Windows

After the installation is completed successfully, open the Windows command prompt and type sbt. Ideally, it should show something similar to Figure 13-3.

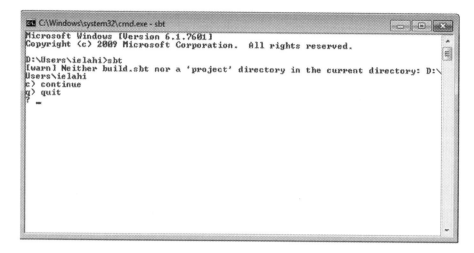

Figure 13-3. Evidence of successful SBT installation on Windows

If you get the message shown in Figure 13-3, it indicates that SBT has been downloaded and installed successfully in your system. Kudos.

Even though we nailed the installation of SBT, we still need to make some arrangements before we can use it. If you parse the message in Figure 13-3, you can see that SBT is complaining about something called a build.sbt (and of project as well but let's ignore that for now). This means that SBT requires some entity known as build.sbt for its operation. So let's focus on that.

Build.sbt for SBT

What is build.sbt? You may have gathered by now that it's something that SBT requires. At a higher level:

- build.sbt is plain text file.

- You as the developer create build.sbt every time you need to work with SBT.

- You use build.sbt for a number of purposes, including dependency management:

- Defining which version of Scala you intend to use (even though you have Scala installed in your system, with SBT you can work with different versions of Scala at the same time for different projects).

- Defining which additional libraries you want to use in your project (Apache Spark will be a library that you will indicate in `build.sbt` going forward).

Where do we find those dependencies? These libraries are generally available on platforms called *repositories*. There are some standard and well-known repositories that developers use a lot and one of those is maven (`https://mvnrepository.com/`).

The libraries or dependencies that you want to use in your program will most probably be available in maven. SBT manages the lifecycle of accessing maven to download those dependencies and making them available for your project. In the world of Scala and Java, those dependencies exist in the form of JARs. It downloads the JARS corresponding to the dependencies that you define in `build.sbt`, from maven to your local system.

I'll highlight more use cases of `build.sbt` going forward, but armed with this knowledge, let's proceed with a working example to put everything in perspective, shall we?

We installed SBT. We need `build.sbt` because we'll use it for dependency management mainly and later for build purposes.

Do the following to create the `build.sbt` file:

1. Create a directory on your system. Call it whatever you want.

2. Open the text editor of your choice, create a new file, and then write the following content in that file as it is. Save the file as `build.sbt` in the folder that you created:

```
name := "HelloWorld"

version := "1.0"

scalaVersion := "2.12.0"
```

Figure 13-4 shows how my folder looks after following these two steps.

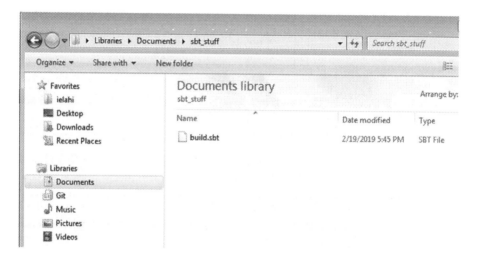

Figure 13-4. *Highlighting the build.sbt file in the project folder*

As you can see, there is only one file build.sbt there. Navigate your cmd to that directory and then type sbt, as shown in Figure 13-5.

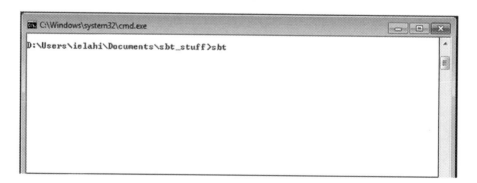

Figure 13-5. *Running SBT from the Windows Command Prompt*

As soon as you press Enter, a lot of stuff will get displayed in your screen, as shown in Figure 13-6.

Figure 13-6. *SBT downloading dependencies from the online repositories (by default, the maven and Scala ones)*

Think of that stuff as the initialization process of SBT. Comparing this to the previous response of SBT, which you got when you didn't have build.sbt in your folder, this one is quite different and verbose.

If you sift through this output, you will find that SBT is downloading myriad JAR files and, as it is downloading them, it's marking them as SUCCESSFUL. Basically, it's downloading all the required dependencies from the default repositories (maven is generally the default repository for SBT). You still may have some questions at this stage, but hold them for a moment and proceed to the next steps.

After it has downloaded all that it needs to download and the process is complete, you will be presented with a shell, like the one shown in Figure 13-7.

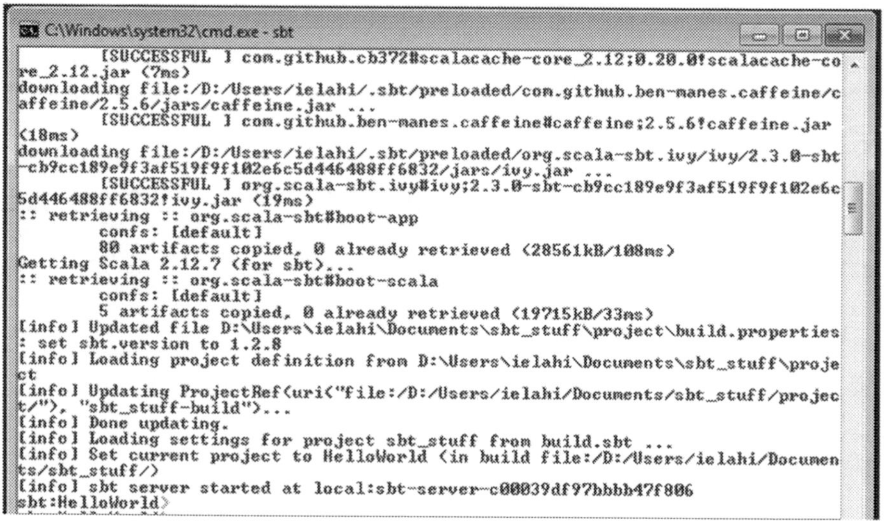

Figure 13-7. *The SBT shell*

This is the SBT shell. This is not the Scala REPL shell. So if you'll start typing Scala commands, you will face disappointment.

This SBT shell can accept SBT-specific commands. We'll use some of these in a while. But to get a feeling, try typing help and it will display all the commands that you can issue here. Similarly, type about and it will display some information as well. It may come as a surprise to you, but you can use the Scala shell/REPL here as well in this context. To do so, type console in the SBT shell and it will launch the good old Scala REPL for you. Upon launching the console the first time, it will also download any required dependencies as specified in the SBT file. Why you would do that will become clearer sooner. See the example in Figure 13-8.

```
sbt:HelloWorld> console
[info] Non-compiled module 'compiler-bridge_2.12' for Scala 2.12.0. Compiling...
[info]   Compilation completed in 16.042s.
[info] Starting scala interpreter...
Welcome to Scala 2.12.0 (Java HotSpot(TM) 64-Bit Server VM, Java 1.8.0_151).
Type in expressions for evaluation. Or try :help.

scala>

scala> println("hello there")
hello there

scala>
```

***Figure 13-8.** Launching the Scala shell from within SBT*

So let's dial back and understand what we actually did:

- You created the build.sbt file and in that build.sbt file, you specified some properties. Mainly, with the following content that you used in build.sbt:

  name := "HelloWorld"

  version := "1.0"

  scalaVersion := "2.12.0"

What you meant was—my project's name is going to be HelloWorld. The same project name appeared in the SBT shell as well.

- The project's version is 1.0. This is for your own reference. Generally you use it when you compile JARs and you can use this property to indicate different versions of JAR.

- Most importantly, you declared a version of Scala. You said that you want to use version 2.12.0 of Scala. This version can be different from the version that you previously installed in your system. Think of this as an independent environment where you can run different versions of Scala and it won't interfere with your installed version. This is a great facility if you want to test library compatibility. Some libraries are available for a specific version of Scala and you can control the behavior using this feature.

- This Scala version is a dependency. You didn't define any additional library or module. You just said that you want a particular version of Scala for your project (the space in your directory is like a project).

- With the build.sbt file created, when you triggered SBT, SBT identified which dependency you mentioned in build.sbt, which was Scala 2.12.0, and it started downloading this dependency from the Internet. Generally, a dependency may depend on other libraries and those libraries may in turn depend on others, so it's like a chain reaction. It's the job of SBT to track all those, download them, and make them available to you. That's what happened when you saw a lot of verbose output upon entering the sbt command at the Command Prompt.

- Once it was done, it launched the SBT shell, which you can use for a number of tasks. But you specifically launched the console, which launched Scala REPL. Did you notice the Scala version when you launched the console?

The image in Figure 13-9 is for your reference.

```
sbt:HelloWorld> console
[info] Non-compiled module 'compiler-bridge_2.12' for Scala 2.12.0. Compiling...

[info]   Compilation completed in 16.042s.
[info] Starting scala interpreter...
Welcome to Scala 2.12.0 (Java HotSpot(TM) 64-Bit Server VM, Java 1.8.0_151).
Type in expressions for evaluation. Or try :help.

scala>

scala> println("hello there")
hello there

scala>
```

Figure 13-9. *Highlighting the Scala version in the Scala shell launched via SBT*

Can you spot Scala 2.12.0 in Figure 13-9? That's because you mentioned that in `build.sbt`. Thus, it managed this dependency and provisioned it for your use. Coolness, isn't it?

Managing Dependencies Using SBT

Before we revert to our original goal, let's take a look at one more aspect. With the console opened, issue the following command:

```
import spray.json._
```

You will be greeted with an error, as shown in Figure 13-10. What happened? You tried to import a library that was not available in your project and Scala was not able to find it.

```
scala> import spray.json._
<console>:11: error: not found: value spray
       import spray.json._
              ^
scala>
```

Figure 13-10. *Importing a library in your Scala session*

Let's say that you want this library in your project. By default, it won't be available for your use. This is where SBT can come to your rescue.

Go to the maven repository and type `spray json` in the search box, as shown in Figure 13-11.

Figure 13-11. *Searching for your library in the maven repository*

In the displayed search results, select Spray JSON. It will take you to the page shown in Figure 13-12.

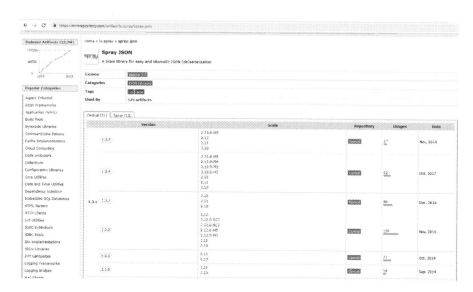

Figure 13-12. *Different versions of the library available on the maven repo*

The versions of this library along with Scala versions are highlighted. Which Scala version did you specify in your `build.sbt` file? It was 2.12.0 so look for the library's version corresponding to this Scala version and select it. There can be overlaps (e.g., 1.3.3, 1.3.4, 1.3.5 all support Scala 2.12). You can choose any of them. Generally, the most recent versions are good and stable (but not always though), so pick any one of them for now. As an example, I picked 1.3.4 and clicked on that. It will take you to the page shown in Figure 13-13.

Figure 13-13. _Specific version of a library on the maven repo, which also highlights how to highlight this as a dependency in different build tools_

On this page, I want you to focus on the section where Maven, Gradle, and SBT are written. All of these are build tools. Maven and Gradle are generally used with Java. SBT is used with Scala, although they can be used interchangeably as well. I selected the SBT tab there and it displayed this text:

```
// https://mvnrepository.com/artifact/io.spray/spray-json
libraryDependencies += "io.spray" %% "spray-json" % "1.3.4"
```

As you can gather, the first line is a comment. Now copy this into your original build.sbt file. Exit the SBT shell prior to that please.

Now my build.sbt file looks like Figure 13-14.

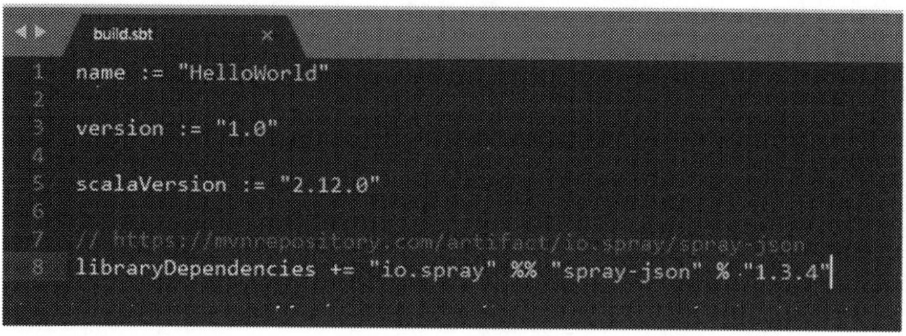

```
◄ ►    build.sbt              ×
1    name := "HelloWorld"
2
3    version := "1.0"
4
5    scalaVersion := "2.12.0"
6
7    // https://mvnrepository.com/artifact/io.spray/spray-json
8    libraryDependencies += "io.spray" %% "spray-json" % "1.3.4"
```

Figure 13-14. *View of the build.sbt file after adding spray JSON as a dependency*

As you can see, I just copy and pasted the libraryDependencies from maven into build.sbt. What I am trying to do is define this as a dependency. I need this library in my project, thus I defined it in my build.sbt and the idea is that when I launch SBT next time, it should manage and provision the dependency for my use. Let's see.

Now when I launch sbt and issue console for instance and try to import this library, it works like a charm, as highlighted in Figure 13-15.

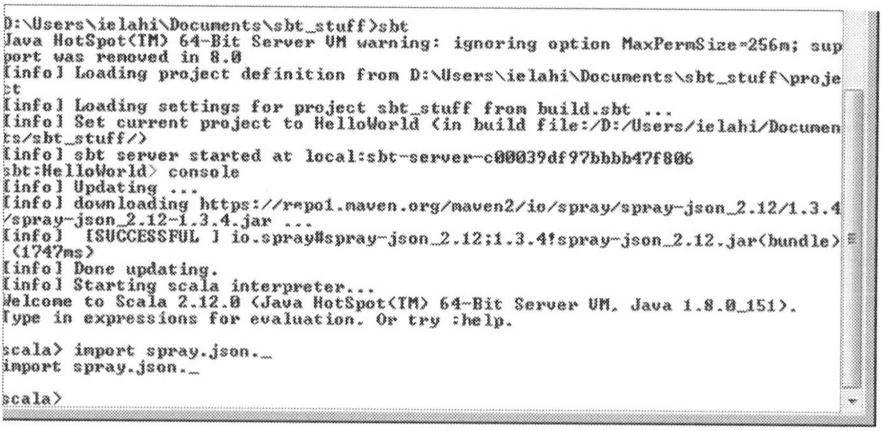

```
D:\Users\ielahi\Documents\sbt_stuff>sbt
Java HotSpot(TM) 64-Bit Server VM warning: ignoring option MaxPermSize=256m; sup
port was removed in 8.0
[info] Loading project definition from D:\Users\ielahi\Documents\sbt_stuff\proje
ct
[info] Loading settings for project sbt_stuff from build.sbt ...
[info] Set current project to HelloWorld (in build file:/D:/Users/ielahi/Documen
ts/sbt_stuff/)
[info] sbt server started at local:sbt-server-c00039df97bbbb47f806
sbt:HelloWorld> console
[info] Updating ...
[info] downloading https://repo1.maven.org/maven2/io/spray/spray-json_2.12/1.3.4
/spray-json_2.12-1.3.4.jar ...
[info]   [SUCCESSFUL ] io.spray#spray-json_2.12;1.3.4!spray-json_2.12.jar(bundle)
 (1747ms)
[info] Done updating.
[info] Starting scala interpreter...
Welcome to Scala 2.12.0 (Java HotSpot(TM) 64-Bit Server VM, Java 1.8.0_151).
Type in expressions for evaluation. Or try :help.

scala> import spray.json._
import spray.json._

scala>
```

Figure 13-15. *Successfully importing a library in a Scala session after it's managed in the build.sbt file*

It actually downloaded the Spray JSON JAR from Maven for us. Why? Because we specified it as a dependency for our project and it obediently grabbed it and provisioned it for us.

This is how you manage dependencies via SBT.

Creating an Executable Scala Application Using SBT

Armed with this knowledge, let's revert to our original goal—creating an executable JAR in Scala.

We will approach that problem in two phases: First, we will understand what it entails to create an executable application and then we'll see how we can package that in the form of a JAR (along with highlighting issues relating to creating Fat JARs).

Let's start simple, yeah? Create a new file in the same folder, put the following content in it, and then save it as `HelloWorld.scala`.

```
println("hello world")
```

Then relaunch `sbt` in the same folder (you can still issue the `run` command without relaunching SBT). Once it's launched, type `run` there.

Once it's done, you will encounter the error shown in Figure 13-16.

```
D:\Users\ielahi\Documents\sbt_stuff>sbt
Java HotSpot(TM) 64-Bit Server VM warning: ignoring option MaxPermSize=256m; sup
port was removed in 8.0
[info] Loading project definition from D:\Users\ielahi\Documents\sbt_stuff\proje
ct
[info] Loading settings for project sbt_stuff from build.sbt ...
[info] Set current project to HelloWorld (in build file:/D:/Users/ielahi/Documen
ts/sbt_stuff/)
[info] sbt server started at local:sbt-server-c00039df97bbbb47f806
sbt:HelloWorld> run
[info] Compiling 1 Scala source to D:\Users\ielahi\Documents\sbt_stuff\target\sc
ala-2.12\classes ...
[   ] D:\Users\ielahi\Documents\sbt_stuff\HelloWorld.scala:1:1: expected class
 or object definition
[   ] println("hello world")
[   ] ^
[   ] one error found
[   ] (Compile /               ) Compilation failed
[   ] Total time: 3 s, completed Feb 19, 2019 6:52:08 PM
sbt:HelloWorld> _
```

Figure 13-16. *Running an executable Scala application that isn't structured properly*

So clearly this doesn't work. Here's what you tried to do:

- Create an executable Scala application that you can run via SBT. You tried to print Hello World on the screen (which you could do without any issues in the Scala shell) and when you tried to run this, it failed in SBT.

- This is because when you want an application to be executable, you have to structure it accordingly. If you go through the error message in this figure, you will find that it says that it "expected class or object definition".

So, how can you write programs that are executable or runnable? To do so, you need three things:

- An object definition (a singleton object)

- A main function with a specific signature

- The body that you want to be executed within the main function

Here's how you do this:

```scala
object HelloWorld {
    def main(args:Array[String]):Unit = {
        println("hello world")
    }
}
```

Pay special attention to the main function. It has to be like this. It has to be called main. It should accept the parameter of type Array[String] and it should return Unit. If you don't follow these steps exactly, you will run into errors.

When you write your program in this way and ask SBT to run, the SBT process will look for an object with the main function that matches the signature. Once it's found, it will mark the entry point of your program execution and your program will start executing from there. Your program will become executable.

Now save the file with this code and type run in the SBT shell. Figure 13-17 shows what happens.

```
D:\Users\ielahi\Documents\sbt_stuff>sbt
Java HotSpot(TM) 64-Bit Server VM warning: ignoring option MaxPermSize=256m; sup
port was removed in 8.0
[info] Loading project definition from D:\Users\ielahi\Documents\sbt_stuff\proje
ct
[info] Loading settings for project sbt_stuff from build.sbt ...
[info] Set current project to HelloWorld (in build file:/D:/Users/ielahi/Documen
ts/sbt_stuff/)
[info] sbt server started at local:sbt-server-c00039df97bbbb47f806
sbt:HelloWorld> run
[info] Compiling 1 Scala source to D:\Users\ielahi\Documents\sbt_stuff\target\sc
ala-2.12\classes ...
[info] Done compiling.
[info] Packaging D:\Users\ielahi\Documents\sbt_stuff\target\scala-2.12\helloworl
d_2.12-1.0.jar ...
[info] Done packaging.
[info] Running HelloWorld
hello world
[success] Total time: 8 s, completed Feb 19, 2019 6:54:50 PM
sbt:HelloWorld>
```

Figure 13-17. *Successfully executing a runnable/executable Scala application*

Success! You were able to successfully display Hello World via an executable Scala application. *Executable* means that your program has an object with the main method, as this serves as an entry point when you run/execute the program. Kudos.

Using the Scala App Trait for Executable Scala Applications

Scala, being a stylish language by all means, provides another way to make your objects executable. You can do so by extending App in your object. It's an OOP concept whereby you make your object inherit from the App trait. But for your understanding, you can think of this as another way to

241

make your Scala program executable and you can avoid defining the main function like before.

Here's an example:

```
object HelloWorld extends App{
    println("hello world")
}
```

Go ahead and try running it in SBT. It should work and everything in the HelloWorld object will run.

Maven Folder Structure for Scala Applications

Maven is a repository. We all know by now. But maven is also a build tool for Java (and for Scala). When developers use maven, they create a specific folder structure and put their code in those folders. SBT also tends to prefer such folder structure as well. This becomes important when you want to organize your classes files in the form of packages (especially in Java, but Scala is still lenient in this context) and when you want to run unit test cases.

Generally, the following folder structure is preferred:

```
project
    build.sbt
    src
        main
            scala
                com
                    irfan
                        elahi
                            HelloWorld
                            HelloFacebook
            resources
```

```
        lib

    test
        scala
            com
                irfan
                    elahi
                        HelloWorldSpec
                        HelloFacebookSpec

        resources
```

As you can see, the code is placed in the `project/src/main/scala` directory. If you want to structure your classes in the form of packages, by convention a folder is created for each coordinate of the package (e.g., for package `com.irfan.elahi`, three folders will be created—`com`, `irfan`, and `elahi`). Just a note that this is not a strict requirement for Scala and is more of a convention.

The `resources` folder is used if you want to include files in the main JAR. In the `lib` folder, you can put JARs as well, which will automatically be added to your classpath.

Now equipped with this knowledge, let's create the directory structure and put our `HelloWorld` class in the `com.irfan.elahi` package.

As per the proposed folder structure, here's how my environment looks:

Figure 13-18. *Example of folder structure for a Scala project highlighting the usage of package (com.irfan.elahi)*

I've modified HelloWorld.scala as follows:

```
package com.irfan.elahi

object HelloWorld extends App{
    println("hello world")
}
```

Upon running it in the SBT shell, it works fine, as shown in Figure 13-19.

```
D:\Users\ielahi\Documents\sbt_stuff>sbt
Java HotSpot(TM) 64-Bit Server UM warning: ignoring option MaxPermSize=256m; sup
port was removed in 8.0
[info] Loading project definition from D:\Users\ielahi\Documents\sbt_stuff\proje
ct
[info] Loading settings for project sbt_stuff from build.sbt ...
[info] Set current project to HelloWorld (in build file:/D:/Users/ielahi/Documen
ts/sbt_stuff/)
[info] sbt server started at local:sbt-server-c00039df97bbbb47f806
sbt:HelloWorld> run
[info] Packaging D:\Users\ielahi\Documents\sbt_stuff\target\scala-2.12\helloworl
d_2.12-1.0.jar ...
[info] Done packaging.
[info] Running com.irfan.elahi.HelloWorld
hello world
[success] Total time: 1 s, completed Feb 19, 2019 7:16:57 PM
sbt:HelloWorld>
```

Figure 13-19. *Successfully executing a Scala application structured as per the maven folder structure*

Also notice that when I run in SBT, it says it's running com.irfan. elahi.HelloWorld. Thus it has identified my package and the naming hierarchy of my package correctly.

Creating Multiple Classes in Your Scala Application and Using Them

Let's do one more thing. Let's create another class and use its function in our main class.

In your existing project, create another file named GreetWorld.scala and put the following content in it:

```
package com.irfan.elahi

object GreetWorld {
        def printMessage(theMessage:String):Unit = {
            println(s"${theMessage} from Irfan Elahi")
        }
}
```

Alter your `HelloWorld.scala` so that it contains the following code:

```
package com.irfan.elahi
import com.irfan.elahi.GreetWorld

object HelloWorld extends App {
        GreetWorld.printMessage("Hello Awesome World")
}
```

Then issue the run command in the SBT shell. You should get the output shown in Figure 13-20. Let's try that first and then look at what's happening in the code snippets.

```
sbt:HelloWorld> run
[info] Compiling 1 Scala source to D:\Users\ielahi\Documents\sbt_stuff\target\sc
ala-2.12\classes ...
[warn] D:\Users\ielahi\Documents\sbt_stuff\src\main\scala\com\irfan\elahi\HelloW
orld.scala:2:24: imported 'GreetWorld' is permanently hidden by definition of ob
ject GreetWorld in package elahi
[warn] import com.irfan.elahi.GreetWorld
[warn]                        ^
[warn] one warning found
[info] Done compiling.
[info] Packaging D:\Users\ielahi\Documents\sbt_stuff\target\scala-2.12\helloworl
d_2.12-1.0.jar ...
[info] Done packaging.
Hello Awesome World from Irfan Elahi
[success] Total time: 1 s, completed Feb 20, 2019 6:07:54 PM
sbt:HelloWorld>
```

Figure 13-20. *Running a Scala application with multiple classes in a package*

As you can see in Figure 13-20, the code runs successfully (with a few warnings that I'll address next).

Here's what we achieved in these code snippets:

- We created two classes in our project (HelloWorld.
  scala and GreetWorld.scala).

- We made HelloWorld.scala an executable class (by
  extending App) in our object.

- We made both the classes belong to the same package,
  called com.irfan.elahi.

- In the GreetWorld.scala file, we created a singleton
  object and defined a function that takes a String
  argument and returns Unit. In the body, the function
  uses the passed parameter and displays a custom
  message on the screen.

- We imported that object into HelloWorld.scala and
  called that object and its function

- And then we ran it.

You should now see:

- How you can create class files in your project and group
  them under one package.

- How you can use elements of one class in another.

- How you can create an executable application.

Pretty cool, isn't it? All of the Scala applications, no matter how
complex and including Apache Spark ones, are based on this foundational
notion. You create classes and use them in one another to achieve the
desired goal. This fosters modularity and reusability of code.

Note Just a callout that we got warnings when we ran the code in SBT. Those warnings were due to the `import` statement that we used in our `HelloWorld.scala` and it's because the usage is redundant in this context. I added that to make the concept of importing classes to your programs easier to understand. Both of the classes belong to the same package and exist at the same hierarchy—`com.irfan.elahi`—thus, in such scenarios, you don't need to explicitly import them.

Compiling Your Scala Applications

At times, you'll want to ensure that there are no syntax errors or any type-related errors in your code. These and many other types of errors can be checked at compile time. Compilation is another step in the overall development lifecycle that, among other things, can be used to quickly perform syntax checks on your code. It takes your `.scala` files, checks for any errors, and then converts them into Java bytecode files (`.class` files), which are then used in later stages to run your applications. When you run your code via SBT, compilation also happens. At times when you don't want to run the code (or your package may not have any executable class at all), the `compile` function helps.

So how do you compile your code via SBT? It's quite straightforward. Just type `compile` in the SBT and it will result in the compilation of your code, as shown in Figure 13-21.

```
D:\Users\ielahi\Documents\sbt_stuff>sbt
Java HotSpot(TM) 64-Bit Server VM warning: ignoring option MaxPermSize=256m; sup
port was removed in 8.0
[info] Loading project definition from D:\Users\ielahi\Documents\sbt_stuff\proje
ct
[info] Loading settings for project sbt_stuff from build.sbt ...
[info] Set current project to HelloWorld (in build file:/D:/Users/ielahi/Documen
ts/sbt_stuff/)
[info] sbt server started at local:sbt-server-c00039df97bbbb47f806
sbt:HelloWorld> compile
[success] Total time: 1 s, completed Feb 21, 2019 9:04:31 AM
sbt:HelloWorld>
```

Figure 13-21. *Compiling a Scala application*

Packaging Scala Applications in the Form of JARs

So far you have learned how to structure your Scala application in the form of folders, packages, and classes and how to compile and run your code using SBT. But if you want to ship your code for execution, you need the code to be packaged. For instance, when you download and install software on the Windows operating system, it's packaged in the form an .exe file that you can use. Such a notion exists in Scala applications as well and Scala applications are generally packaged in the form of JAR files.

On a similar note, when you write Apache Spark programs using its APIs in Scala, you will package your applications in the form of JAR files and will run them on the *cluster,* which refers to multiple computer systems/servers configured in a way to run distributed processing workloads on them.

So how do you generate JAR files from your Scala code? That's where SBT comes to the rescue again! Just like the commands you used before—like run and compile—there is another command, called package, that you can use to package your Scala applications in the form of JAR files. When you run the package command, it compiles it as well so you don't need to separately run the compile command prior to package.

Refer to Figure 13-22 for further clarity.

```
sbt:HelloWorld> compile
[success] Total time: 1 s, completed Feb 21, 2019 9:04:31 AM
sbt:HelloWorld> package
[success] Total time: 1 s, completed Feb 21, 2019 9:09:26 AM
sbt:HelloWorld>
```

Figure 13-22. *Packaging a Scala application in the form of a JAR*

After you issue this command, SBT will package your code in the form of a JAR and it will be deployed in the following folder in your project:

```
Project_folder\target\scala-2.12\helloworld_2.12-1.0.jar
```

The exact folder names and class names can vary depending on the version of Scala that you specified in SBT, the name of your project, and the version of your project, but it will be in this same hierarchy.

Once you have the JAR, you can execute using the following command outside of SBT:

```
scala <location_of_jar_file>
```

Refer to Figure 13-23.

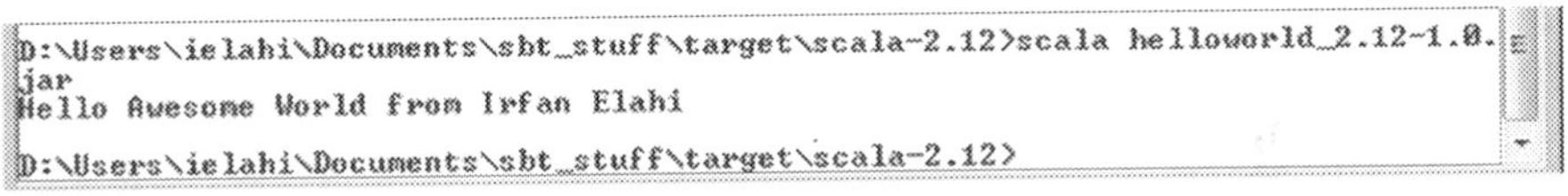

```
D:\Users\ielahi\Documents\sbt_stuff\target\scala-2.12>scala helloworld_2.12-1.0.
jar
Hello Awesome World from Irfan Elahi
D:\Users\ielahi\Documents\sbt_stuff\target\scala-2.12>
```

Figure 13-23. *Running a packaged JAR file*

Once it's done, the the program will start execution from the class where you used the main function (or extended the App trait).

Transitioning to an IDE

You've taken a long journey so far. You've learned about Scala, the Scala REPL, and then graduated to using SBT to run your Scala applications. But the road to excellence still lies ahead.

You need to align yourself to the practices of professional developers. If you do so, your productivity will be significantly boosted. There will be a small learning curve, but it's all worth it. So which tools do the professional developers use for their development? You will often hear the term IDE from them, which stands for Integrated Development Environment. It's software where developers write their code and use for a number of reasons. The IDE provides syntax checking, highlighting, type-checking, dependency management, and build, compile, and unit test execution, all within one platform. So it significantly improves productivity.

First, you wrote your code in Scala REPL and then in a text editor and now is the time to start using an IDE.

Every language has its go-to IDEs that are considered the de facto ones. For .NET development, it's Visual Studio. For Python, it's PyCharm (though this is purely subjective and preference can vary from one person to another). For Java and Scala, it's IntelliJ IDEA (many developers also love Eclipse a lot, but I prefer IntelliJ IDEA). Again, it's a matter of personal preference.

So in this chapter and from now on, we will pivot to IntelliJ.

Installing IntelliJ IDEA

Before you can use IntelliJ, you need to download and install it. To do so, go to this website:

```
https://www.jetbrains.com/idea/download/
```

You will find a page similar to Figure 13-24.

Figure 13-24. *Downloading IntelliJ*

You will find two options to download IntelliJ IDEA. For many purposes, the Community Edition works fine and is free. So just download the Community Edition for now.

Once it's downloaded, install it. The installation process will be a no brainer. Once it's installed, launch IntelliJ.

When you launch IntelliJ IDEA, you will see something like what's shown in Figure 13-25.

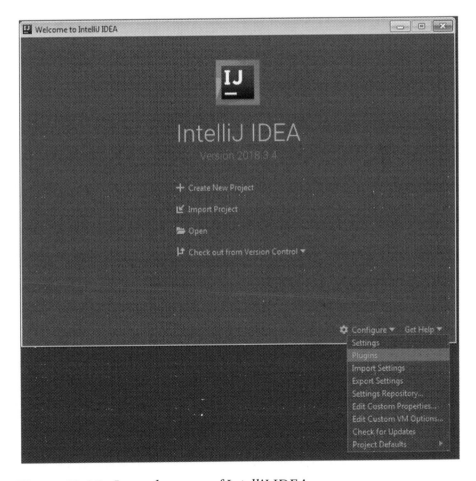

Figure 13-25. *Launch screen of IntelliJ IDEA*

IntelliJ IDEA Plugins Installation

When you install IntelliJ IDEA, it doesn't come with much functionality that is required for Scala development. You therefore need to install a set of plugins. At a bare minimum for Scala development, you need the following two plugins:

- Scala

- SBT

You can install the plugins by selecting Plugins, as shown in Figure 13-25, and then typing the required plugin names in the search box. Then you install them, as shown in Figure 13-26.

Figure 13-26.   *Installing plugins in IntelliJ IDEA*

As you install plugins, it will probably ask you to restart IntelliJ IDEA. Please do so.

Importing a Project in IntelliJ IDEA

You have already been working on a project in this chapter where you created two classes (HelloWorld and GreetWorld) and compiled and executed them via the SBT shell.

Let's import that project and do the same in IntelliJ IDEA.

1. Launch IntelliJ and click Import Project (refer back to Figure 13-25).

2. Navigate to the directory in the IntelliJ IDEA wizard where your project exists.

3. Click Next and then select Import Project From External Model. Select sbt, as shown in Figure 13-27.

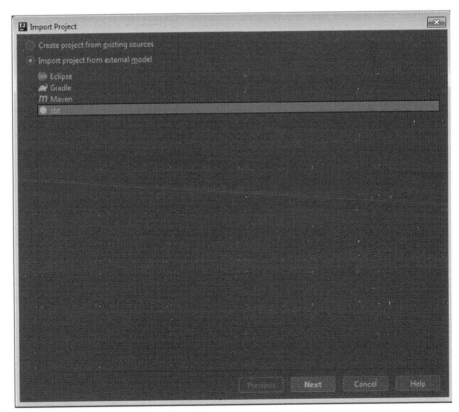

Figure 13-27. Selecting SBT in IntelliJ IDEA

4. In the next screen, ensure that JDK is selected. If not, click New and navigate to the directory where you installed JDK on your system.

5. Also, click the Global SBT settings and then the select Custom in Launcher (`sbt-launch.jar`) section. Navigate to the directory where you installed SBT. In that directory, go to `bin` and select `sbt-launch.jar`. You can use bundled SBT that comes with IntelliJ but it helps to use the one that's installed in your system. This helps when you have done some configurations in your installed SBT (e.g., proxy settings). Refer to Figure 13-28.

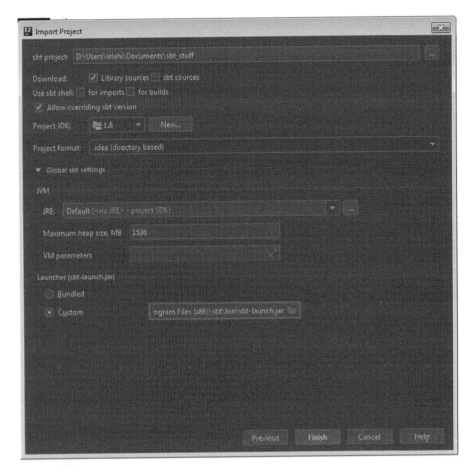

Figure 13-28. *Configuring JDK and SBT in IntelliJ IDEA*

6. Click Finish.

Once it's done, allow IntelliJ IDEA to finish the import. At this time, it will parse the build.sbt file and download any dependencies that weren't downloaded before (by the way, when the dependencies are downloaded, they exist on your system and are managed by Apache Ivy).

Once IntelliJ IDEA has done its work and imported all the dependencies, you can do the development in this environment.

Figure 13-29 shows a view of that state.

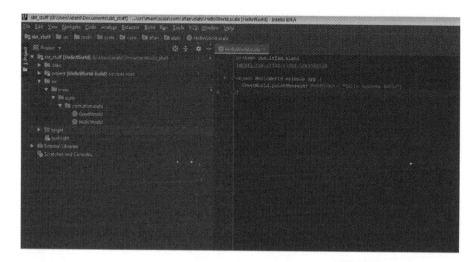

Figure 13-29. *View of IntelliJ IDEA after it has successfully imported a project*

As you can see in Figure 13-29:

- The left pane shows the directory structure of your project.

- The right pane shows the contents of the files that you selected in the left pane. I selected `HelloWorld.scala` and it's displayed there.

You can also see that it has beautifully highlighted the syntax in the right pane. Each keyword in your code is highlighted with a color (e.g., the `object` and `extends` keywords).

It also highlights when something is wrong. For instance, as mentioned, the `import` command is redundant and it has been highlighted in gray and is underlined. Similarly, if you make any mistakes, it will show in real time what you are doing, which really helps achieve prompt troubleshooting.

It also shows which elements are available in classes that you are using. For example, refer to Figure 13-30.

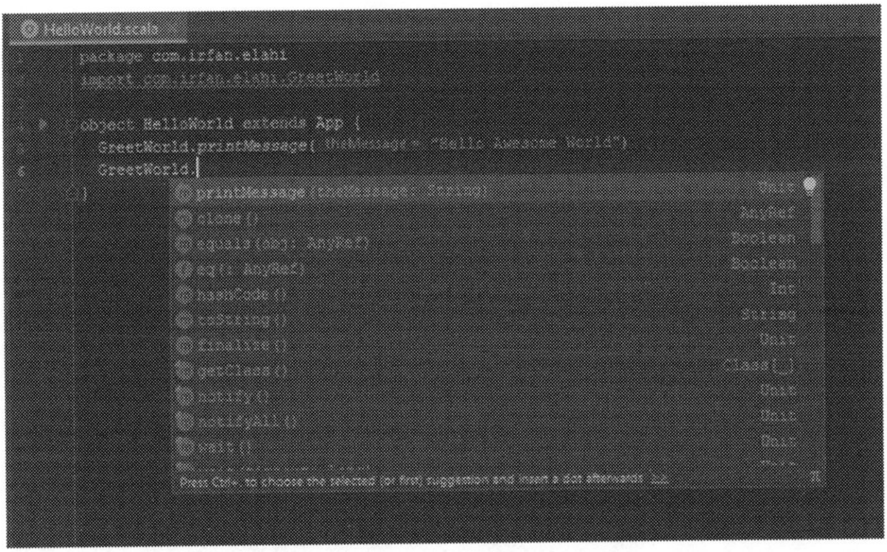

Figure 13-30. *Highlighting elements of classes/objects*

As you can see in Figure 13-30, it shows that `printMessage` function is available in the `GreetWorld` object and shows which parameters this function accepts and the value it returns. Pretty handy!

Now let's run our Scala application. You can do so from within IntelliJ IDEA. Right-click the class that you want to run (in this case, `HelloWorld.scala`) from the left pane and then click Run 'HelloWorld', as shown in Figure 13-31.

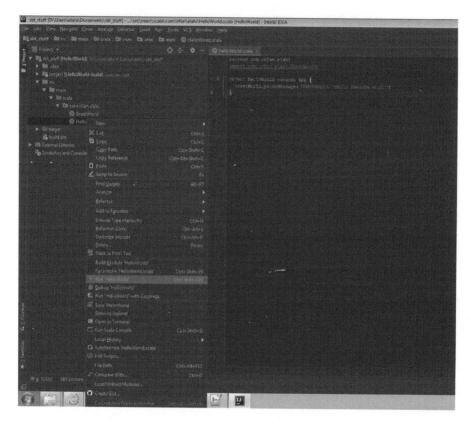

Figure 13-31. Running an application from within IntelliJ IDEA

As you click that, it will start the whole process (compile and run) at the back end and will show the output at the bottom of IntelliJ IDEA, as shown in Figure 13-32.

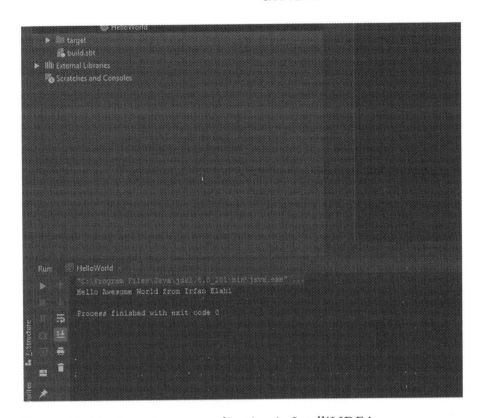

Figure 13-32. *Running an application in IntelliJ IDEA*

Since IntelliJ IDEA is a sophisticated tool with a lot of amazing features, I can't possibly cover everything. But it's a great idea to familiarize yourself as much as possible with its features and learn how you can use them to enhance your productivity.

In this chapter, we covered a lot. You started using SBT for different tasks like compile, run, and package and you learned how to structure your Scala applications and how to use IntelliJ IDEA to supercharge your development efforts. This lays a substantial foundation for the upcoming chapter, where we use Apache Spark while employing all the concepts we've learned so far! Exciting times!

EXERCISES

- Create a new project in IntelliJ IDEA instead of importing one.

- Install SBT in a Linux machine and use it.

- Figure out how to change the configurations of SBT. For example, if you are behind a network proxy, how do you specify those settings in SBT, how do you change the heap size of SBT, etc.

- Find out the meaning of "uber" or "Fat" JAR.

- Understand the different constructs of `build.sbt` and how to include different repositories.

- Figure out how to specify multiple dependencies in `build.sbt`.

- Figure out how to define multiple projects in `build.sbt` and how to structure your folders accordingly. Also, how do you selectively build/package individual projects in SBT.

- Use maven to build Scala code instead of SBT. Understand the differences between the two.

- Instead of specifying the required dependencies in `build.sbt`, find alternative ways to provide those dependencies for your project. (Hint: It involves putting JAR files in your project's folder.)

- Figure out how to launch the Scala shell and add different JARs to its classpath.

CHAPTER 14

Hello Apache Spark

Doesn't it feel good when you are in the vicinity of your envisioned and cherished destination? When you see in retrospect that you've been through a long journey and the milestone that you once dreamed of is in your reach? You must have the same feeling as you start this chapter, because this last chapter of the book is all about how you can put the concepts you've learned into practice and work on Big Data analytics and Apache Spark. So buckle up as it's going to be an exciting ride!

I've referred to Apache Spark again and again in this book. I shed some light to some extent on this technology in the first chapter of the book. But let's formally develop some basic understanding of Apache Spark and then see how you can start doing development in Scala.

Revisiting Apache Spark

There are many ways I can explain Apache Spark and I already covered it some in the first chapter. Let's develop a better understanding of it.

Distributed Computing Engine

Apache Spark is a distributed computing engine. What this means is that Spark can run computations on multiple machines. Normally, if you write a Scala program and run it, all the processing within the program happens on a single machine. It's not like multiple machines participate in the computation. But in Spark, you have the opportunity to leverage

© Irfan Elahi 2019
I. Elahi, *Scala Programming for Big Data Analytics*,
https://doi.org/10.1007/978-1-4842-4810-2_14

pools of machines to perform distributed computation. This is the crux of Big Data—horizontally scaling out to multiple machines to perform analysis on data that can't be done in one machine. A single machine has limits to scalability, in that you can increase processing power (CPU cores or multiple processors) or memory (RAM) or storage (hard disk) only to some extent. But with the horizontal scalability model, there is practically no limit to the amount of data you can process (and store). This is where Spark shines—when writing programs that result in execution on multiple machines.

Spark and Hadoop

Spark is strongly associated with Hadoop. Hadoop, as mentioned in the first chapter, is a consortium of services that fall in different categories:

- Compute (Spark, MapReduce, Hive, Impala, Storm, Flink, Samza, Drill, and Presto)

- Storage (HDFS, Kudu, HBase, MongoDB, Cassandra, and Gremlin)

- Security (Sentry, Knox, and Ranger)

- Metadata Management (Hive Metastore, HCatalog, Atlas, and Cloudera Navigator)

- Message Queues (Kafka, EventHub, and Kinesis)

- Integration (NiFi, StreamSets, and Flume)

- Cluster Manager (YARN and Mesos)

This list is by no means exhaustive. But you get the idea. Spark is a compute service of Hadoop. Generally, the Hadoop ecosystem is deployed on a cluster of machines. Some of the services run on some nodes. So is the case with Apache Spark. As a part of cluster configuration, it's configured where Spark processes will run.

Spark and YARN

YARN, as you can see in the previous list, is a cluster manager. Cluster managers play a crucial role in the overall anatomy of the cluster. In a cluster, each machine has some compute resources (processor cores and RAM). There has to be a component that has a broader visibility of these compute resources and their respective status. The component should know where to launch a particular process—on which machine—provided that machine has resources available to accommodate those resources. The component should know that if multiple jobs are being submitted to the cluster, how it manages the workload—queue them or throttle the resources consumed by each job. Also, if a distributed process is running and if that process gets killed on a node, it should know where to spawn that process next. All of these and many other jobs are the responsibilities of YARN.

Like many other components, YARN has two components:

- Resource Manager

- Node Manager

The Resource Manager is a master process and the Node Managers are like slave processes that run on multiple machines. YARN allocates compute resources via something called a *container* (which is an abstraction for compute resources). Thus, the YARN Resource Manager launches a container on the Node Managers and it tracks the status of containers (health, failure, etc.).

You may wonder why we are discussing YARN. It's because in production-grade systems, Spark is run on top of YARN. When a Spark job runs, Spark processes run on nodes running Node Manager processes (though this is not universally true as Databricks Spark offering doesn't run on YARN per se). Spark jobs get resources from YARN in the form of containers. Understanding this will help you in the long run.

Spark Processes

Spark consists of two types of processes:

- Driver processes

- Executor processes

A Driver process runs in something called a driver JVM. Think of it as master process for Spark. This is responsible for coordinating the flow of Spark tasks. This is where Spark Context is initialized. There is always one driver process per Spark application. It doesn't participate in the distributed processing of tasks. All of the distributed processing happens in the executor processes. Also, when you load a file or any data in Spark, it gets partitioned and loaded in the heap memory of the Executor processes' JVM.

Spark Abstractions

You may have heard the term RDD before in this book. Now is the time to develop some understanding of what RDD is. It stands for Resilient Distributed Datasets. The best way to understand RDD is:

- It is like a Scala collection in that it provides a number of methods like `map`, `filter`, and `foreach`. They are used in the same way as in a Scala collection.

- However, unlike Scala collections, they don't exist on one machine or JVM. They are partitioned. They exist on multiple nodes (specifically executor processes that run on YARN Node Managers).

- They are immutable. You know pretty well by now what that means.

- They are resilient. Spark keeps track of RDD lineage— how they were created. So if any RDD gets destroyed, Spark knows how to recreate them.

How you create RDDs? There are multiple ways. You can create them when you load data from external systems to Spark for processing (e.g., filesystems, databases, and message queues). You can create them by parallelizing Scala collections as well (which doesn't make sense as there is little rationale to distribute a collection that can already fit in a driver JVM but is used for testing purposes). Or you can transform one RDD to create others.

We'll work with RDDs in a bit later in this chapter.

Lazy Execution Model of Spark

One of the most interesting features of Spark is that it follows the lazy execution model. If you invoke certain functions (called transformations) on RDDs, the execution won't happen right away. Rather, such transformations are "recorded" in the form of DAGs (directed acyclic graphs, a kind of graph that means that it doesn't form a circular loop). For example, if you specify the following set of transformations on an RDD:

- Load data from a filesystem

- Filter those lines that have the word "Scala" in them

- Split each line on a comma

- Count how many such rows exist

Then the first three transformations won't result in execution. Rather Spark will start recording those steps in the form of a graph like this:

```
Load Data ➤ Filter ➤ Map
```

As soon you issue certain methods (called actions), it will trigger execution from the start of the stated graph, i.e., loading data from the filesystem.

This is called the *lazy execution model* and Spark relies on it to perform optimizations under the hood.

I hope that this provides a very basic and foundational understanding of Apache Spark. For deeper understanding, I suggest you research on your own and to facilitate that further, I recommend you enroll in my best-selling Udemy course. It will take your Apache Spark skills to the next level:

```
https://www.udemy.com/apache-spark-hands-on-course-big-data-
analytics/
```

Enough of the theory. Let's do some actual Spark development, shall we?

First, we'll do ad hoc Spark development and then we'll see how to create a Spark application that we can run on a cluster (you won't exactly need a cluster to follow along because we will simulate one).

Apache Spark in Scala in Action

So how can you get your hands dirty with Apache Spark? There are so many ways to go about it. But the quickest and free way to get started with using Apache Spark is via:

- Cloudera QuickStart VM

- Databricks Community Edition

Both Cloudera and Databricks are commercial vendors of Hadoop. Databricks is solely based on Apache Spark though. With the first option, you can run a Linux VM on your Windows system that has Cloudera Distribution of Hadoop (CDH) installed and configured in it. Spark also comes with CDH. In there, you can launch `spark-shell` (or `spark2-shell`), which will launch Scala Shell, with which you are quite familiar by now. It has all the dependencies of Apache Spark preconfigured.

But if you don't want to go through the hassle of running VM in your system, the other option is to use Databricks. It's free and it runs in a browser, which means you need to be connected to the Internet to use it. Another great feature of Databricks out of the box is that it provides Notebooks UI, which is quite handy for ad hoc analysis.

Spark Environment Setup in Databricks

Let's use Databricks! To do that, you need to sign up for Databricks Community Edition. Go to the following link:

```
https://databricks.com/try-databricks
```

Refer to Figure 14-1.

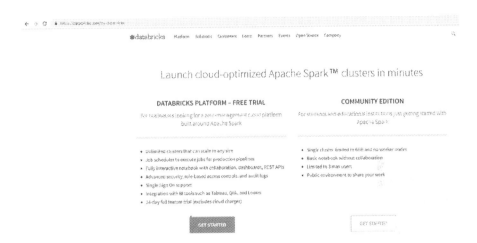

Figure 14-1. *Signing up for Databricks Community Edition*

Click on GET STARTED below the Community Edition and follow the signup process. Once the signup process is complete, you will be brought to the screen shown in Figure 14-2.

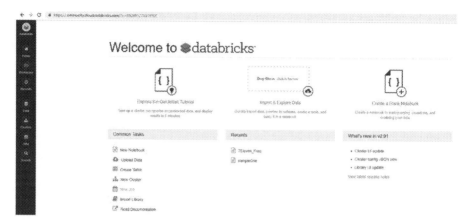

Figure 14-2. *Main screen of Databricks after signing in*

Before you can use Apache Spark, you need to have a cluster up and running. The Community Edition of Databricks provides such a facility. Though the cluster that can be set up in the Community Edition is quite minimal and single node, it will be enough to serve our requirements.

Thus, go to the Clusters option on the menu on the left side. Once there, click the Create Cluster button. See Figure 14-3.

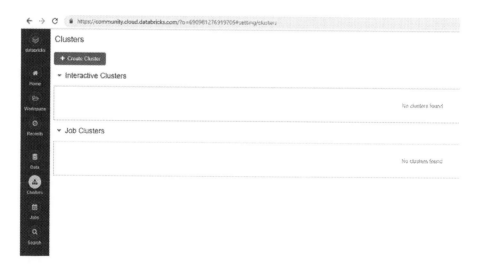

Figure 14-3. *Cluster section of Databricks*

Once you click the button, it will ask you a couple of options related to creating clusters, as shown in Figure 14-4. Specify any cluster name that you like and select the Scala version of your choice. Then click the Create Cluster button. It will take a minute or so for the cluster to get up and running. You can track its status in the Clusters section.

Figure 14-4. *Cluster creation options in Databricks*

Once the cluster is created, click Databricks in the left menu and click New Notebook. It will ask for name of the notebook, which language you want to use, and which cluster to use. Specify any name for the notebook, choose Scala as the language, and select the cluster that you just created. Finally, click Create. Once it's done, it will launch a notebook for you in all of its glory!

Refer to Figure 14-5.

Figure 14-5. *A notebook in Databricks*

A notebook consists of different cells where you write your code and the output gets displayed in the section right below the cell. You can create cells above and below the current cell and work in a highly interactive manner. Data scientists use notebooks a lot.

In each cell, you can write Scala expressions and press Ctrl+Enter to execute that cell. If you press Shift+Enter, it will execute the cell and create another cell below it.

So with that knowledge in mind, let's do some Apache Spark programming using Scala. You will find that the concepts that you have learned so far will be quite handy.

Again, as this book is not about Apache Spark itself, I'll not go into the nitty-gritty Spark-specific details. If you want to quench your curiosity, you can always research on your own. (Though there are a number of books on Apache Spark, I think the one by Butch Quinto published by Apress (link: https://www.apress.com/gp/book/9781484231463) is particularly good as it not only covers Spark but also talks about its integration with other Big Data technologies for different use cases. I know this because I was the technical reviewer of that book.)

Apache Spark Development in Scala

Now when you work with Apache Spark, the first thing that you do is initialize the SparkContext object. It's an object that provides an entry point to the Spark cluster. This object exposes a number of functions that you can use and via which you use the distributed computation capability of Apache Spark. From the Scala standpoint, SparkContext is just an object, like any other that you create from a class. The good thing about Databricks is that when you launch notebook, the SparkContext object is already made available to you so the overhead of creating one is removed. However, when we'll write our Scala applications in the next section, we'll actually create one.

In a Databricks notebook, as mentioned, the SparkContext object is made available as the sc variable, as shown in Figure 14-6.

Figure 14-6.   Spark Context object

As you can see, just like in Scala REPL, it gives you in the output the type of object as well. You can see that the sc object is of type org.apache. spark.SparkContext.

Let's see what elements are available in the sc object. You can do so by typing sc. and pressing Tab. Just like in Scala REPL.

Now let's create an RDD. As mentioned, there are multiple ways to create RDDs. The simplest way is via the sc.parallelize method. This method accepts a Scala collection (with which you have worked before). This program will distribute the contents of a single node Scala collection, which you've been using, to multiple nodes (though in this example, you are working with single node cluster).

So type this:

```
sc.parallelize(List(1,2,3,4,5,6,7,8,9,10))
```

and press Ctrl+Enter or Shift+Enter, as shown in Figure 14-7.

```
Cmd 1
  1  sc.parallelize(List(1,2,3,4,5,6,7,8,9,10))
res3: org.apache.spark.rdd.RDD[Int] = ParallelCollectionRDD[0] at parallelize at command-3199234926143690:1
```

Figure 14-7. *Using the parallelize function of Spark Context object*

You will find that upon execution, you will get an object. That object will be of type `org.apache.spark.rdd.RDD[Int]`. It's just a type. Just like any other types that you have worked with before.

You can assign the created RDD to an immutable variable:

```
val myRdd = sc.parallelize(List(1,2,3,4,5,6,7,8,9,10))
```

Then you can work with this variable.

Now let's say you want to multiply each element of the RDD by 2. How will you do that? Before you answer, think about how will you do that if you were operating with a list. In fact let's start with a list and then see how we can do that in RDD:

Here's how you address such a problem when you are working with a list:

```
val aList = List(1,2,3,4,5,6,7,8,9,10)
aList.map(x=>x*2)
```

It should be pretty clear to you now. You used `map` function of list and then used a function (`x=>x*2`) that operated on each element of the list and multiplied each by 2.

How will you do that in RDD? Answer: In exactly the same way!

```
myRdd.map(x=>x*2)
```

However, as mentioned, there are two types of functions in Spark: transformations and actions. Here you used the map function of RDD. It's a transformation. When you'll execute this command, you won't see any output. In fact, Spark hasn't executed anything owing to its lazy execution model. So how do you trigger execution? You use an action. One of the actions is collect, which returns all the elements of an RDD to the driver JVM. (It's highly recommended to *not* use collect when you are working with large-scale datasets because doing so will cause all executor processes to return their part of RDD data to the driver program and thus it will cause memory overflow problems for the driver program. (This is not just because you are currently using the Community Edition of Databricks. It's a general issue that can appear because the driver program is a single JVM process with limited memory so if you try to bring data from multiple executor JVM processes into a single JVM, it's bound to experience memory overflow problems.)

As we are working with a very small dataset, it's safe to use it for now. Developers tend to use collect because doing so converts RDD to Scala collections (localized in the driver JVM) with which they are already familiar. However, you can (and should) use Spark's functions like map, filter, etc. to do any type of required transformation and you will be reaping the benefits of distributed computing while doing so.

Thus, when you use collect, you get the output as follows:

```
myRdd.map(x=>x*2).collect()
```

Refer to Figure 14-8.

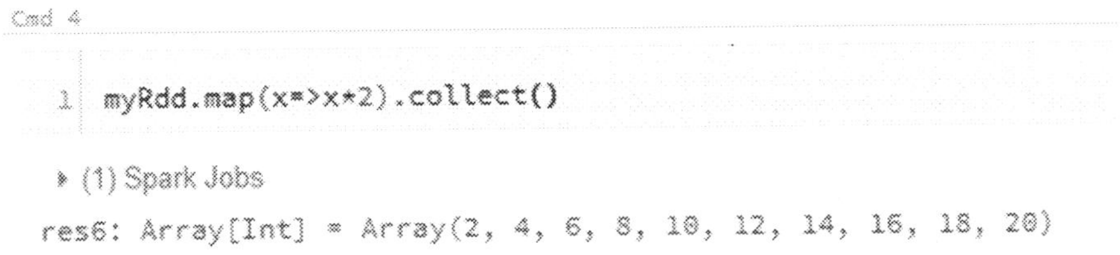

```
Cmd 4

1  myRdd.map(x=>x*2).collect()

▶ (1) Spark Jobs
res6: Array[Int] = Array(2, 4, 6, 8, 10, 12, 14, 16, 18, 20)
```

Figure 14-8. *Using the collect() action on RDD*

And there you have it. You can use the same logic whether you are operating on a small dataset or a massive dataset comprised of billions of records. If you use this process, it will distribute the processing on a cluster of machines for you. Spark takes away all of those complexities and presents a very simple RDD API model that significantly resembles Scala collections.

Similarly, you can use the `filter` function on the RDDs. Let's filter just the even numbers from the list. Also, let's define a function that does the job as well to see that your concepts about functions will work perfectly fine here too:

```
def giveEvens(givenNumber:Int):Boolean = {
  givenNumber%2 == 0
}

myRdd.filter(x=>giveEvens(x)).collect()
```

This is shown in Figure 14-9.

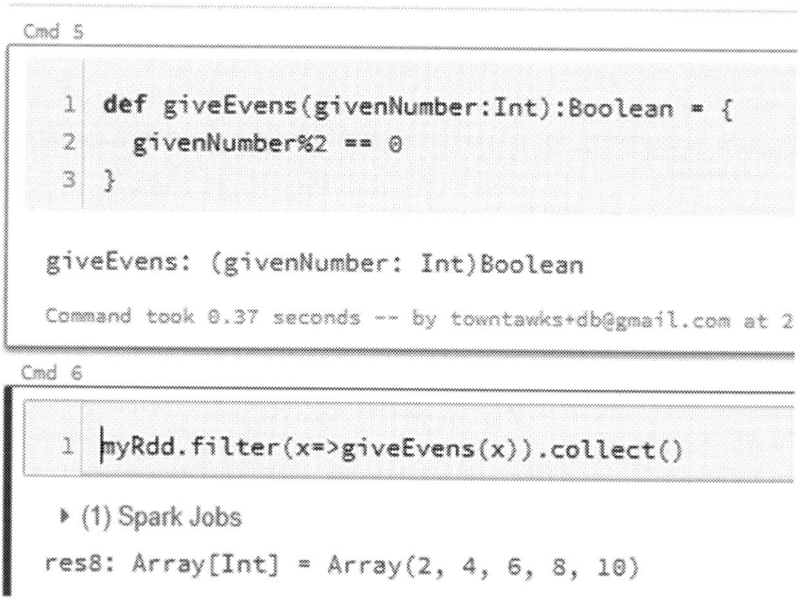

```
Cmd 5

1  def giveEvens(givenNumber:Int):Boolean = {
2      givenNumber%2 == 0
3  }

giveEvens: (givenNumber: Int)Boolean
Command took 0.37 seconds -- by towntawks+db@gmail.com at 2

Cmd 6

1  myRdd.filter(x=>giveEvens(x)).collect()

▶ (1) Spark Jobs
res8: Array[Int] = Array(2, 4, 6, 8, 10)
```

Figure 14-9. *Using a function in Spark's filter transformation*

There you go! Here's a quick recap of what you did:

- You defined a function that takes an integer number as an argument and checks if its modulus by 2 is zero (i.e., when divided by 2, whether the remainder is zero).

- You used the `filter` function of RDDs and used that function there. Filter is also a transformation so to actually trigger execution and to get results, you used the `collect()` action, which caused Spark to perform the processing and return the results (in this case, a list of even numbers).

You can chain the operations together instead of writing them separately in each expression:

```
myRdd.filter(x=>giveEvens(x)).map(x=>x*2)
```

You can also seamlessly use collection functions on the results that get returned when you use the `collect()` action. The `collect()` function returns the contents of RDD as an array and thus you can use the `foreach` function of arrays to further do any processing on them (such as print them):

```
myRdd.filter(x=>giveEvens(x)).map(x=>x*2).collect().foreach(println)
```

Note that `foreach` in this context is used on `Array` and not on RDDs. Many RDD functions and collections functions have the same name, so it's important to be aware of the context in which you are using them.

EXERCISES

Explore available transformations and actions of Apache Spark RDD APIs. For example, transformations: `foreach`, `reduce` and `zipWithIndex` and actions: `collect`, `action`, `take`, and `first`. Use them in your code. You can use them without any barrier. You have all the required knowledge of Scala and Spark to do that. So trust yourself and do that.

Converting an RDD to a Dataframe

Apache Spark programming itself deserves its own book, but I wanted to cover another related concept where you'll see the use of case classes in action as well. So far you have used RDDs of Apache Spark. RDDs are the fundamental data abstraction of Apache Spark and they provide collections like interfaces. But many times when you are interacting with structured or tabular data, Spark provides another abstraction that facilitates the processing logic significantly: the dataframe.

If you have used Python's pandas or R, the similar notion of dataframe data structure exists and it resembles a table. A dataframe consists of rows and columns and each column has a name and type. This data structure provides its own set of functions, which are optimally suited for relational data processing. You can select a few columns, create new columns, join two dataframes, filter certain rows, or apply a function to one or more columns. As per Apache Spark, the direction is to emphasize dataframes and that's the reason that Spark Machine Learning libraries and Spark streaming libraries are now tailored to use Spark SQL dataframes. Both of them had RDD APIs as well, but the trend is going in favor of the Spark SQL dataframe.

If you have an RDD, you can convert it to dataframe. You can create a dataframe a number of ways, including by loading CSV, JSON, XML, and Parquet files into Spark, and reading from databases, to name two. But to emphasize the use of case classes and to connect the dots, I'll show the method where you convert an RDD to dataframe while using case classes. This method is called *Schema reflection* in the Apache Spark literature.

Uploading Data to Databricks

First, let's create a small CSV file and upload it to Databricks. We'll use that as an example. This example resembles a lot of real-world scenarios in which you get data in the form of files and process them in Apache Spark.

In your local system, create a simple CSV file (I called it sample_file. csv) in a text editor of your choice and write a few records, as follows:

```
1,irfan,scala,pakistan
2,mark,java,usa
3,ehsan,python,portugal
4,arslan,java,pakistan
```

It's a simple CSV file (i.e., each field is delimited or separated by a comma) that has four rows and four columns in it.

Let's upload sample_file.csv to Databricks. To do so, click Data in the left menu and then click Add Data, as shown in Figure 14-10.

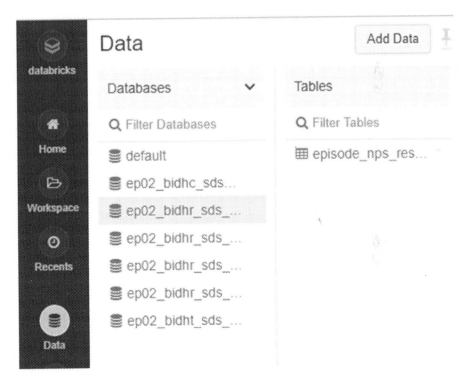

Figure 14-10. *Uploading data on Databricks*

Once you're done, click on Browse, as shown in Figure 14-11.

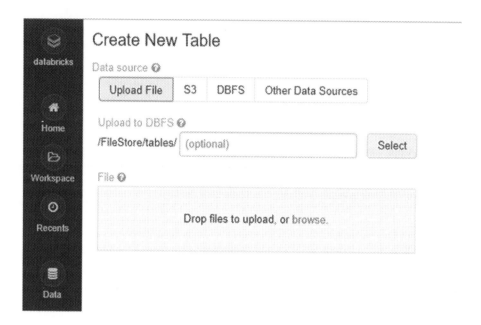

Figure 14-11. *The Browse option for file to be uploaded on Databricks*

Select the file from your local filesystem. Once it is uploaded, it will look like Figure 14-12.

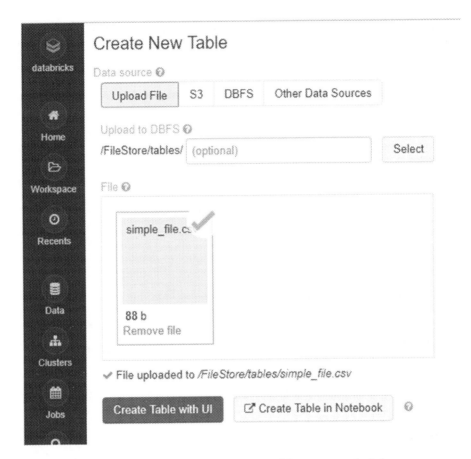

Figure 14-12. Successfully uploading a file on Databricks

Note the location where the file is uploaded (`/FileStore/tables/simple_file.csv` in this case) and create a new notebook.

Converting an RDD to a Dataframe

You first need to load the data in Apache Spark so that you can perform some analysis on it. In Spark, there are many ways to do that, but let's try the following way:

```
val myData = sc.textFile("/FileStore/tables/simple_file.csv")
```

279

What this will do is load the contents of the text file in an RDD. As you can see, we used sc (SparkContext) here. We used its function (.textFile) and passed a String argument to it (representing the path of the file that you just uploaded). Try to understand the syntax in the way I explained previously and everything will make sense. Using this function, you will get an RDD. Just like each list has a type (representing the type of elements it holds), so is the case with RDD. When you use the sc.textFile function, you get RDD[String], i.e., the RDD of type String.

Now let's create a case class that will represent the schema or structure of our dataframe. We need to define the name and type of each column. You already know how to create case classes, and the same concept will be applicable here:

```
case class DataSchema(id:Int, name:String, language:String,
country:String)
```

I defined a case class with the name DataSchema and defined four attributes in it, which represent the schema of my file in this context.

You need to process your RDD. Your RDD is currently an RDD[String], i.e., each element is a string. You need to split each String element into Array[String] by splitting on the comma , character. Wherever it finds a comma, it will split the String. You have already used these functions.

You can do that by using RDD's map transformation, which will look exactly the same as the collection's map function:

```
myData.map(x=>x.split(","))
```

Then you need to convert each Array[String] element to a DataSchema object. You've done this same activity before as well. Here's how you'll do it in one go:

```
val myDF = myData.map(x=>x.split(",")).map(x=>DataSchema(x(0).toInt,
                                             x(1),
                                             x(2),
                                             x(3)
                                        )).toDF
```

In this code in the second map, you mapped each component of each `Array[String]` to a corresponding case class attribute. Note that you had to type-cast `x(0)` to `Int` because that represents the `id` attribute of the case class, which is `Integer`. After you've done it all, you just need to call the `.toDF` function. As soon as you do, you'll see Spark SQL's powerful dataframe in action! Be sure to save it in a variable as we did in the previous step.

Using the Spark SQL Dataframe API

Let's play around with that dataframe. You can select columns of your dataframe using the `.select` method and passing in a column name as `String` (there is another possible variation in the arguments as well but let's park it):

```
myDF.select("id").show()
```

Refer to Figure 14-13.

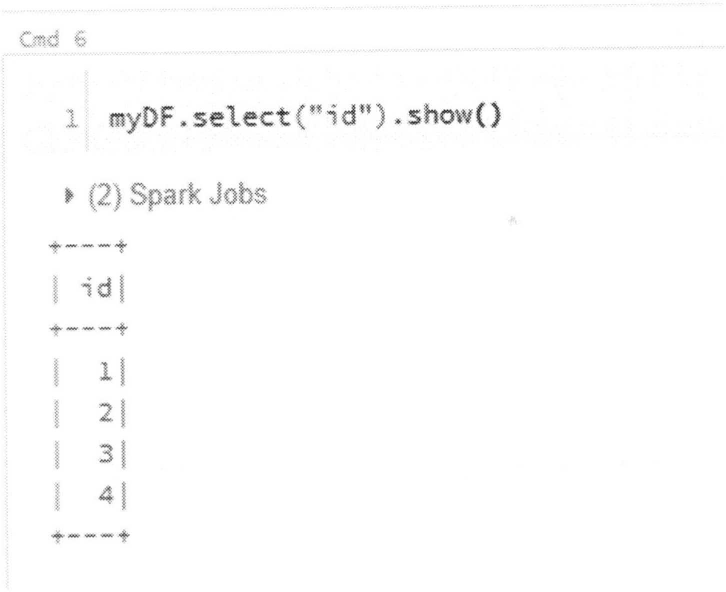

```
Cmd 6

1   myDF.select("id").show()

  ▶ (2) Spark Jobs

  +----+
  | id |
  +----+
  |  1 |
  |  2 |
  |  3 |
  |  4 |
  +----+
```

Figure 14-13. *Using the Select function of the Spark SQL dataframe*

As you can see, we used the `.show()` function here to display the output to the screen. For dataframes, it's an action and you know what an action does.

You can filter rows of this dataframe based on the condition of your choice:

```
myDF.filter("language == 'java'").show()
```

Refer to Figure 14-14.

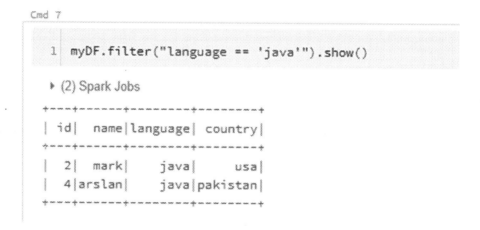

Figure 14-14. *Using the filter function of the Spark SQL dataframe*

As you can see, its APIs are quite intuitive and simple in structure.

Before we conclude, let me highlight another cool feature of Apache Spark. If you know SQL, you may know already that this is one of the most heavily used languages for data analysis and is generally the standard language for querying relational databases. It's a declarative language, which allows you to express your intent and let the system figure how to execute it. It's for this reason that many business intelligence and data analysts use it a lot, as they don't need to learn and use programming languages to explicitly define the logic to do the processing.

Running SQL Queries Against Spark SQL Dataframes

If you have a Spark SQL dataframe, you can, to everyone's amazement, run SQL queries against it! It's a highly powerful feature of Apache Spark and is one of the main drivers of its massive adoption in enterprises all around the world.

From the Scala standpoint, you need to use another object, called spark (which is a SparkSession object), and use its .sql function to pass SQL queries as strings. Before that, you need to issue another function that registers that dataframe as a table (and that table exists just within that session).

```
myDF.createOrReplaceTempView("df_temp")
```

So with this function, you created an alias of your dataframe that can be queried via SQL.

Next, use the spark.sql function to issue queries against this table:

```
spark.sql("select * from df_temp where country = 'pakistan'").show()
```

Refer to Figure 14-15.

Figure 14-15. *Using SQL commands against the Spark SQL dataframe*

I'll conclude this section at this point. So far, you've developed some exposure of some APIs available in Apache Spark (RDD and dataframe) and were able to use their respective functions to perform Big Data analytics. Let me reiterate: the same logic and constructs are used whether you are using a single node or multi-node cluster or operating on smaller or larger datasets.

Now let's look at how you can use the concepts related to SBT and OOP to create a Spark Application that you can run in a cluster.

Creating Spark Applications Using SBT

Just like creating any other Scala applications, the workflow of creating Spark application is almost the same:

1. You structure your project's folder as per maven standards (having code in `src/main/scala` folders and test cases in `src/test/scala`).

2. You use `build.sbt` and define dependencies (in this case, you define Apache Spark related dependencies).

3. You use SBT to package the code in the form of a JAR.

However, there are some caveats and I'll highlight them as we go along.

Creating a New Project in IntelliJ IDEA

Let's get started:

1. Launch IntelliJ IDEA and choose New Project.

2. Select Scala on the left and SBT in the right pane, as shown in Figure 14-16.

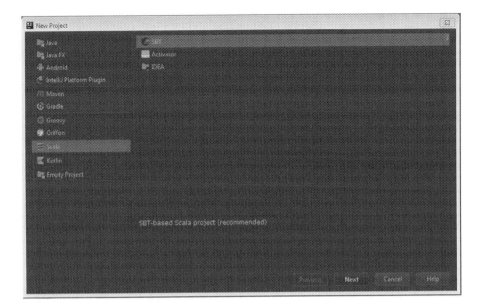

Figure 14-16. *Creating a new project in IntelliJ IDEA*

3. Fill out the project-related details like name and
    location of the project directory. Ensure that JDK is
    selected. For this project, choose 2.11.8 as the Scala
    version. For SBT, you can choose 0.13.18 or later and
    click Finish, as shown in Figure 14-17.

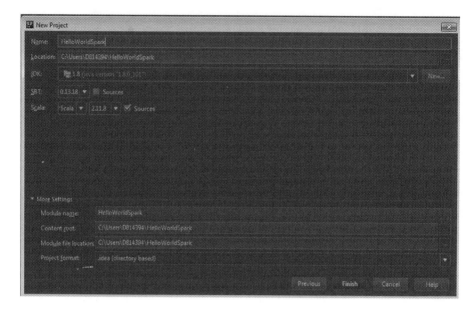

Figure 14-17. *Scala project details in IntelliJ IDEA*

Once it's done, IntelliJ IDEA will do its stuff and grab and manage all the dependencies relevant to your project (mainly SBT and Scala at this stage). So give IntelliJ some time and track its activity in the bottom-most bar.

Once it has done its work, you will find that it has created all the required directories for you, even `build.sbt`! Cool.

Managing SBT Plugins for Uber JARs

Let's double-click `build.sbt` and then populate it as per our requirements of dependencies. Before that, just a callout that you need to manage an SBT plugin so that you get Uber or Fat JARs. Remember these terms in the previous chapter? What they basically mean is that you want to create a JAR that contains all the dependencies that you highlight in `build.sbt`. You want them to be packaged in that JAR and, as you stuff those dependencies in that JAR, your JAR becomes "Fat" or "Uber". It's a reliable way to manage dependencies, although it results in somewhat bulky JARs in terms of size.

The other option is that you manage the dependencies in the environment where you are working, i.e., you configure the Java classpath and Spark so that your program can find the dependencies for you during runtime. This approach can be risky, as not all platforms can be configured per your expectations. Thus I generally rely on building Uber JARs to avoid such surprises.

To do that, you add the SBT Assembly plugin to your project. Adding that is pretty easy. Just create a new file in the `project` folder in your working directory and call it `assembly.sbt`. Then populate it with the following content:

```
addSbtPlugin("com.eed3si9n" % "sbt-assembly" % "0.14.9")
```

That's it. You've added the required plugin that will allow you to build Fat JARs having all dependencies within it.

Managing Apache Spark Dependencies in SBT

Next, you need to specify Apache Spark as dependencies in the `build.sbt` file. The concept is the same as in the previous chapter. However, as Spark is deeply integrated with Hadoop, the dependencies list can vary depending on what you intend to do (e.g., if you want to integrate with HDFS and Hive, then you will have to include those dependencies as well). I'll keep matters simple here, so the content of `build.sbt` will look like this:

```
name := "HelloWorldSpark"

version := "1.0"

scalaVersion := "2.11.8"

libraryDependencies ++= Seq(
```

```
  "org.apache.spark" %% "spark-core" % "2.4.1",
  "org.apache.spark" %% "spark-sql" % "2.4.1"
)

assemblyJarName in assembly := "hello_spark.jar"

assemblyMergeStrategy in assembly := {
  case PathList("META-INF", "MANIFEST.MF") => MergeStrategy.discard
  case x => MergeStrategy.first
}
```

Most of the constructs in build.sbt should look familiar. There are few quirks however:

- Instead of specifying library dependencies individually, I've created a Seq() or collection of them and specified them as a command-separated list.

- I've also added the assemblyJarName section, which is related to the SBT assembly plugin. The JAR that will be created when you'll do the packaging should have this name.

- Then you have the assemblyMergeStrategy section, which is also related to the SBT assembly package. This section is about managing conflicts if two classes have the same name. If that happens, you can instruct SBT assembly plugins how to handle conflicting classes. There are various merge strategies. For example, if such conflicts happen, you can discard, pick one, and so on. Go through the SBT assembly plugin's page to develop a better understanding of this. At this stage, it suffices to say that you are using SBT assembly plugin's construct and have specified merged strategies. These conflicts do appear and that's why I've added them to avoid any issues if you are reproducing this project on your end.

Now with `build.sbt` created and populated, you will find that IntelliJ will again do some processing and will download and index the required dependencies for you. Let it do its work.

Spark Application Code

After that, create a new Scala file in IntelliJ and call it `HelloWorldSpark.scala`. Populate it with the following content:

```
import org.apache.spark.SparkConf
import org.apache.spark.sql.SparkSession

object HelloWorldSpark extends App{

  //initialize Apache Spark Context:
  val conf = new SparkConf()
  val spark = SparkSession.builder.config(conf).getOrCreate()

  val aList = List(1,2,3,4,5,6,7,8,9,10)
  val aRdd = spark.sparkContext.parallelize(aList)

  aRdd.filter(x=>x%2==0).map(x=>x*2).collect().foreach(println)

}
```

Here's what we did in this code snippet:

1. You imported required classes from the Spark package. You are able to do so because you specified Apache Spark as a dependency in `build.sbt`.

2. You created an executable Scala application by using an object and extending `App`.

3. You created an object called `conf` from `SparkConf` class using the new keyword. You then used that `conf` object to create another object called `spark`.

It's the same SparkSession object that you used in your Databricks notebook. You always create these objects when working with Spark 2 (which is the latest version of Spark to date).

4. You then accessed the sparkContext object from the SparkSession object and used the familiar parallelize function to parallelize a Scala list that you created previously.

5. You then used the same series of transformations (filter to filter just the even numbers) and mapped to double each number. You then called the collect action and used foreach on the returned Scala array to display each element of the array. Simple stuff!

That's it. Your Apache Spark application will filter even numbers from a list and will double them and display them. This is like a HelloWorld for Spark! Figure 14-18 shows how my project looks in IntelliJ.

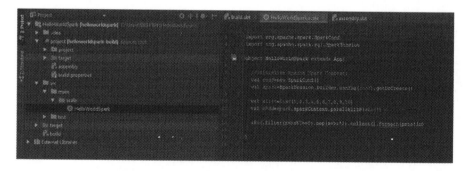

Figure 14-18. *Project view in IntelliJ IDEA*

Now let's compile and package the code. You can launch Terminal from within IntelliJ, as shown in Figure 14-19.

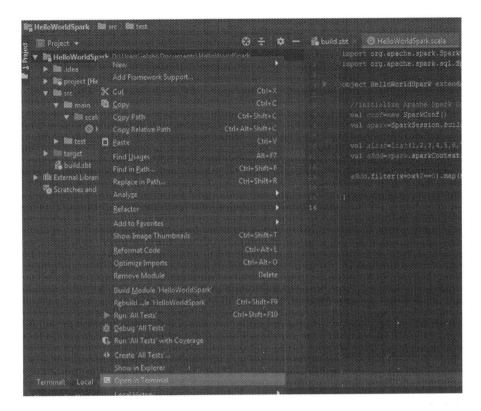

Figure 14-19. *Launching Terminal from IntelliJ IDEA*

Once Terminal is started, launch the SBT shell using the sbt command. Then type compile to compile your project to check if there are any compile-time errors.

If not, type assembly. This will use the SBT assembly plugin to package your Scala code in the form of a Fat JAR. This is the alternative of the package command that you used previously.

Refer to Figure 14-20.

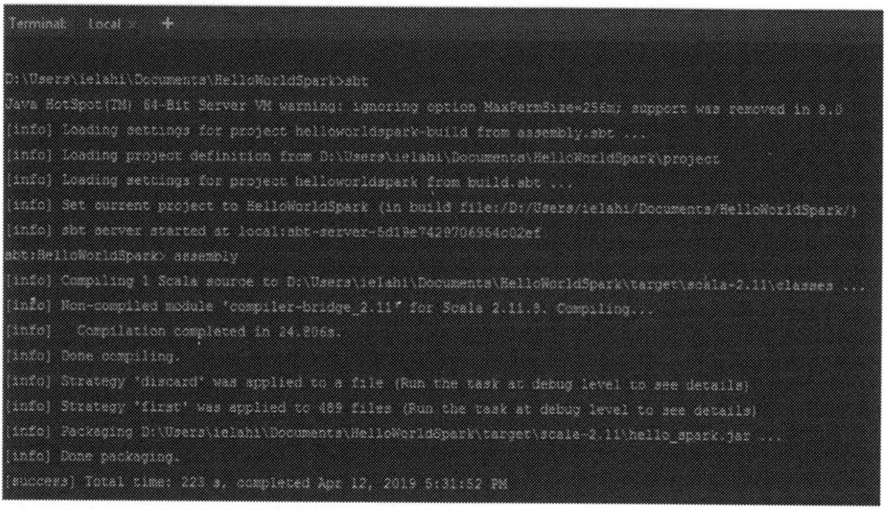

Figure 14-20. *Compiling and packaging Scala code from the SBT shell in IntelliJ IDEA*

The output is further shown here:

```
D:\Users\ielahi\Documents\HelloWorldSpark>sbt
Java HotSpot(TM) 64-Bit Server VM warning: ignoring option
MaxPermSize=256m; support was removed in 8.0
[info] Loading settings for project helloworldspark-build from
assembly.sbt ...
[info] Loading project definition from D:\Users\ielahi\
Documents\HelloWorldSpark\project
[info] Loading settings for project helloworldspark from build.sbt ...
[info] Set current project to HelloWorldSpark (in build
file:/D:/Users/ielahi/Documents/HelloWorldSpark/)
[info] sbt server started at local:sbt-server-5d19e7429706954c02ef
sbt:HelloWorldSpark> assembly
[info] Compiling 1 Scala source to D:\Users\ielahi\Documents\
HelloWorldSpark\target\scala-2.11\classes ...
```

[info] Non-compiled module 'compiler-bridge_2.11' for Scala
2.11.8. Compiling...
[info]   Compilation completed in 24.806s.
[info] Done compiling.
[info] Strategy 'discard' was applied to a file (Run the task
at debug level to see details)
[info] Strategy 'first' was applied to 489 files (Run the task
at debug level to see details)
[info] Packaging D:\Users\ielahi\Documents\HelloWorldSpark\
target\scala-2.11\hello_spark.jar ...
[info] Done packaging.

Once the code issued, it will create a JAR for you and will store it in
target/scala-2.11/location, as shown in Figure 14-21.

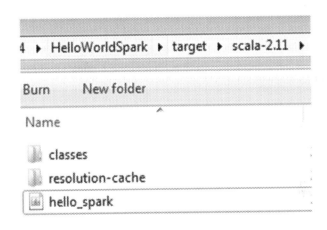

Figure 14-21. *The JAR created after packaging/assembling in SBT*

Congratulations. Your JAR is ready in all of its splendor.

Next, you need to run this JAR or Spark application. For that, you need
to have a Spark cluster running. Unfortunately, you can't use Databricks
Community Edition. To simulate such an environment, use Cloudera
QuickStart VM. The process of setting it up should be straightforward.

First, download the Oracle VirtualBox software from the Internet and download Cloudera QuickStart VM from Cloudera's website (a few Google queries will take you there, don't worry). Then, using VirtualBox, start the QuickStart VM. Once it's started, launch the Linux terminal and you should be good to go.

I am assuming that you have launched Cloudera QuickStart VM. Copy your JAR to that VM. It's time to launch your Spark application via the JAR, which is the final milestone of this book and of your amazing journey. To do so, issue the following command:

```
spark2-submit --class HelloWorldSpark --master local hello_
spark.jar
```

Note that if `spark2-submit` doesn't work because of the change in Cloudera QuickStart versions, try issuing `spark-submit` (Just a note that in case you find encounter errors related to signature of JAR file, you can issue the following command to address that: `zip -d file.jar 'META-INF/*.SF' 'META-INF/*.RSA'`).

Figure 14-22 shows the same command being issued in an environment.

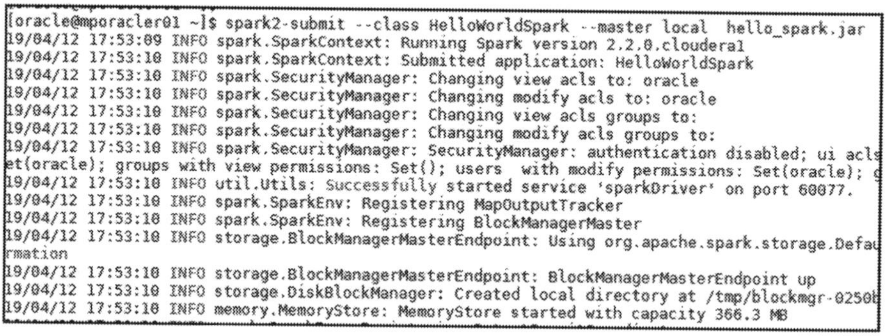

```
[oracle@mporacler01 ~]$ spark2-submit --class HelloWorldSpark --master local hello_spark.jar
19/04/12 17:53:09 INFO spark.SparkContext: Running Spark version 2.2.0.cloudera1
19/04/12 17:53:10 INFO spark.SparkContext: Submitted application: HelloWorldSpark
19/04/12 17:53:10 INFO spark.SecurityManager: Changing view acls to: oracle
19/04/12 17:53:10 INFO spark.SecurityManager: Changing modify acls to: oracle
19/04/12 17:53:10 INFO spark.SecurityManager: Changing view acls groups to:
19/04/12 17:53:10 INFO spark.SecurityManager: Changing modify acls groups to:
19/04/12 17:53:10 INFO spark.SecurityManager: SecurityManager: authentication disabled; ui acls
et(oracle); groups with view permissions: Set(); users  with modify permissions: Set(oracle);
19/04/12 17:53:10 INFO util.Utils: Successfully started service 'sparkDriver' on port 60077.
19/04/12 17:53:10 INFO spark.SparkEnv: Registering MapOutputTracker
19/04/12 17:53:10 INFO spark.SparkEnv: Registering BlockManagerMaster
19/04/12 17:53:10 INFO storage.BlockManagerMasterEndpoint: Using org.apache.spark.storage.Defau
rmation
19/04/12 17:53:10 INFO storage.BlockManagerMasterEndpoint: BlockManagerMasterEndpoint up
19/04/12 17:53:10 INFO storage.DiskBlockManager: Created local directory at /tmp/blockmgr-0250b
19/04/12 17:53:10 INFO memory.MemoryStore: MemoryStore started with capacity 366.3 MB
```

Figure 14-22. *Issuing the spark2-submit command*

Here's a portion of the output:

```
spark2-submit --class HelloWorldSpark --master local  hello_
spark.jar
19/04/12 17:53:09 INFO spark.SparkContext: Running Spark
version 2.2.0.cloudera1
19/04/12 17:53:10 INFO spark.SparkContext: Submitted
application: HelloWorldSpark
19/04/12 17:53:10 INFO spark.SecurityManager: Changing view
acls to: oracle
19/04/12 17:53:10 INFO spark.SecurityManager: Changing modify
acls to: oracle
19/04/12 17:53:10 INFO spark.SecurityManager: Changing view
acls groups to:
19/04/12 17:53:10 INFO spark.SecurityManager: Changing modify
acls groups to:
19/04/12 17:53:10 INFO spark.SecurityManager: SecurityManager:
authentication disabled; ui acls disabled; users  with view
permissions: Set(oracle); groups with view permissions: Set();
users  with modify permissions: Set(oracle); groups with modify
permissions: Set()
19/04/12 17:53:10 INFO util.Utils: Successfully started service
'sparkDriver' on port 60077.
19/04/12 17:53:10 INFO spark.SparkEnv: Registering
MapOutputTracker
19/04/12 17:53:10 INFO spark.SparkEnv: Registering
BlockManagerMaster
19/04/12 17:53:10 INFO storage.BlockManagerMasterEndpoint:
Using org.apache.spark.storage.DefaultTopologyMapper for
getting topology information
19/04/12 17:53:10 INFO storage.BlockManagerMasterEndpoint:
BlockManagerMasterEndpoint up
```

```
19/04/12 17:53:10 INFO storage.DiskBlockManager: Created local
directory at /tmp/blockmgr-0250bd6c-6a3c-4a12-b255-28af82af630f
19/04/12 17:53:10 INFO memory.MemoryStore: MemoryStore started
with capacity 366.3 MB
19/04/12 17:53:10 INFO spark.SparkEnv: Registering
OutputCommitCoordinator
19/04/12 17:53:10 INFO util.log: Logging initialized @1856ms
19/04/12 17:53:10 INFO server.Server: jetty-9.3.z-SNAPSHOT
19/04/12 17:53:10 INFO server.Server: Started @1923ms
```

The output indicates the logs that Spark generates when initializing itself to run the application. You don't need to understand what each of the log output lines means at this stage.

With this command, you launched a Spark application. The syntax for launching Spark applications is via spark2-submit (you don't use scala <jar> in this context). Then you specify the main class using --class parameter. Then you specify --master local to specify the cluster manager where the application should run. I've run it in local mode (i.e., single node), but if you have YARN set up, run it on YARN using the --master yarn option (that's how it's executed in production systems). Then you specify the JAR that you copied over to the VM.

When you issue this command, it will display a whole lot of stuff on the screen. Don't be intimidated by that. Amidst that output, you will find your desired output, as shown in Figure 14-23.

```
19/02/22 10:58:30 INFO client.TransportClientFactory: Successfully created con
200 after 107 ms (75 ms spent in bootstraps)
19/02/22 10:58:30 INFO util.Utils: Fetching spark://
f-4733-9a03-0ebcff16069e/userFiles-e7495967-f71f-456e-a44e-5d856e4ebf9f/fetchF
19/02/22 10:58:30 INFO executor.Executor: Adding file:/tmp/spark-a71431c5-3b7f
e4ebf9f/hello_spark.jar to class loader
19/02/22 10:58:30 INFO executor.Executor: Finished task 0.0 in stage 0.0 (TID
19/02/22 10:58:30 INFO scheduler.TaskSetManager: Finished task 0.0 in stage 0.
19/02/22 10:58:30 INFO scheduler.TaskSchedulerImpl: Removed TaskSet 0.0, whose
19/02/22 10:58:30 INFO scheduler.DAGScheduler: ResultStage 0 (collect at Hello
19/02/22 10:58:30 INFO scheduler.DAGScheduler: Job 0 finished: collect at Hell
4
8
12
16
20
19/02/22 10:58:30 INFO spark.SparkContext: Invoking stop() from shutdown hook
19/02/22 10:58:30 INFO server.AbstractConnector: Stopped Spark@4aa3d36{HTTP/1.
19/02/22 10:58:30 INFO ui.SparkUI: Stopped Spark web UI at http://
19/02/22 10:58:30 INFO spark.MapOutputTrackerMasterEndpoint: MapOutputTrackerM
19/02/22 10:58:30 INFO memory.MemoryStore: MemoryStore cleared
19/02/22 10:58:30 INFO storage.BlockManager: BlockManager stopped
19/02/22 10:58:30 INFO storage.BlockManagerMaster: BlockManagerMaster stopped
19/02/22 10:58:30 INFO scheduler.OutputCommitCoordinator$OutputCommitCoordinat
19/02/22 10:58:30 INFO spark.SparkContext: Successfully stopped SparkContext
19/02/22 10:58:30 INFO util.ShutdownHookManager: Shutdown hook called
```

Figure 14-23. *Output of Spark application*

Can you spot the numbers 4, 8, 12, 16, and 20? Isn't this the desired output of your Spark program? We filtered just the even numbers from a list of numbers (1 to 10) and doubled each number.

So there you go—you were able to create an executable Spark application, compile its Uber JAR, and then run it in a cluster environment successfully. Trust me, this is a huge achievement. Many "experts" or "gurus" in Big Data struggle a lot with such stuff. But here you are—able to apply all the concepts of Scala you've learned (and some limited concepts of Spark) to create a full-blown application. Kudos! I can't emphasize enough how amazing that achievement is.

Conclusion and Beyond

Congratulations on completing this book. It's indeed a commendable milestone and you must pat yourself on the back. However, this is not the end. In fact, it's the start of so many amazing things that you need to further learn, master, and excel in to gain a competitive edge in the world of Big Data.

In my view, if you completed this book, it's quite probable that you are serious about learning Big Data. Big Data is a huge domain on its own with so many dimensions to it. But one of the best ways to get started is to understand and develop skill in at least one of the tools in the Big Data ecosystem while keeping a holistic view of others.

In this context, Apache Spark is the leading Big Data processing engine and should be your focus if you are starting up or even if you are experienced in this domain. As per a recent survey report (see `https://www.techrepublic.com/article/the-top-10-big-data-frameworks-used-in-the-enterprise/`), Apache Spark tops the Big Data framework used by enterprises today.

So what's next for you? The way I see it, there are two ways you can go further:

- *Big Data path:* Start learning Apache Spark. As you learn Apache Spark, you will naturally be exposed to other Hadoop technologies like Hadoop Distributed Filesystem (HDFS), Hive, HBase, Kafka, etc., because of Spark's strong integration with all these. How do you learn Apache Spark? You can get the Apress book that I mentioned or you can enroll in my best-selling Udemy course, which has been featured multiple times as the highest rated course there. You can enroll in my course by following the link:

```
https://www.udemy.com/apache-spark-hands-on-
course-big-data-analytics
```

- *Advanced Scala path*: There is still much to learn
  about Scala. You still have to learn how to use
  Scala's object oriented and functional programming
  constructs in a thorough way, how to develop Scala
  applications using build tools like SBT, and how to do
  test-driven development in Scala. Additionally, Scala
  has a number of strong frameworks like Play (a web
  framework), which the world is leveraging for building
  microservices. Similarly, akka is yet another library for
  building highly concurrent applications. There is not a
  single resource that you can rely on for learning these.
  I am planning to write more books and launch courses
  to cover many of these areas as well, so watch for these
  things happening!

All in all, one becomes an outstanding developer by practice. So
practice as much as you can. Try to write at least one line of code each
day. Work on harder problems and come up with your approach to solving
them. You'll fail but if you learn from your mistakes, this whole journey
will lead you to the excellence that you deserve. All the best! You can
always reach out to me via my blog (`http://www.irfanelahi.com`) or my
LinkedIn profile (`https://linkedin.com/in/irfanelahi`).

It's been an amazing journey with you all. Now go and claim the
excellence that you deserve!

That concludes this book, folks.

Regards,

Irfan Elahi

Index

© Irfan Elahi 2019
I. Elahi, *Scala Programming for Big Data Analytics*,
https://doi.org/10.1007/978-1-4842-4810-2